ROAMING WILD

THE FOUNDING OF COMPASSION IN WORLD FARMING

EMMA SILVERTHORN

Whittles Publishing

For Nan and Peter, who were inspirational
and above all else, superb grandparents

Also to the memory of John Callaghan, former
Compassion employee and wonderful human being

Published by
Whittles Publishing Ltd.,
Dunbeath,
Caithness, KW6 6EG,
Scotland, UK

www.whittlespublishing.com

ISBN 978-184995-543-0

Printed by Short Run Press Ltd

Endorsements for

Roaming Wild

Emma has created an amazing book to celebrate the lives and work of two extraordinary people – the couple who created and founded Compassion in World Farming, a highly respected animal welfare organisation today. In the family story that runs alongside this excellent detailed reporting of their many campaigns, with all their frustrations and successes, and their many distinguished supporters and helpers, we read a touching history of the two people responsible for it all. CWF, founded by Emma's grandfather and grandmother, is something that Emma has grown up with – so thank you, Ems, for sharing its history with us.

And thank you, Compassion, for all you do for animals!

Sue Jameson, actor and animal campaigner

How is it possible for two people to change the world? Find out by reading *Roaming Wild*. Emma Silverthorn tells the compelling story of her grandparents, Peter and Anna Roberts. In 1967 they gave up farming to launch Compassion in World Farming, the pioneering international force for farmed animals, linking animal cruelty with world hunger. *Roaming Wild* inspires compassionate action now for animals, humans and the Earth.

Kim Stallwood, animal rights author and independent scholar

Magnificent! A captivating and heart-warming biography of giants of the animal welfare movement – a timely arrow-of-light for life on earth.

This uplifting story of the battle to free farm animals from cages needed to be told, in order to help humanity free its thinking away from greed towards compassion, choosing a saner world for future generations.

Read it; but beware – you may end up with the question 'Do my actions truly treat all sentient beings with respect?'

Tess Burrows, adventurer, author and motivational speaker.

Roaming Wild by Emma Silverthorn is a terrific read. This is a biography of a relationship made in heaven but firmly rooted on earth, a love not only for each other but for our wonderful planet and all those species with whom we share it, particularly those animals imprisoned in our food systems. It's the story of two amazing people, Anna and Peter Roberts, whose kindness and empathy turned compassion into action and founded Compassion in World Farming. I urge you to read it.

Peter Egan, actor, director and animal rights activist

This is a fascinating and touching book, about an organisation with which I've worked and whose ethos I've loved for many years. Compassion in World Farming is built on respect – and, of course, compassion – for animals. It's clear from this book that the founders, Peter and Anna, were deeply compassionate about the world and the people affected by our conventional farming attitudes, too. This is a very open and honest account of the lives of the Roberts family and the people with whom they worked tirelessly for a better life for farmed animals. I hope that the more people who read about their prescient vision of what intensive farming would do to the world, the faster the change towards a truly better, and more compassionate, world for both animals and people will be.

Dr Emma Milne BVSc FRCVS, vet, author and animal welfare advocate

Contents

Preface

Many years ago Philip Lymbery, CEO of Compassion in World Farming (CIWF), asked if I'd ever thought about writing the life story of my grandparents, Anna and Peter Roberts. It was they who, back in 1967, had been the founders of the animal welfare charity.

A year or so later the CIWF trustees, some of whom had worked with Anna and Peter and had known them well, commissioned me to write the story. I was thrilled to be given the time, space and licence to investigate my family history. Many years of reading, research, interviews, writing and editing, and laughter and frustration followed.

The resultant book, *Roaming Wild*, is part biography and part modern-day animal welfare history; it details the unconventional and pioneering lives of my grandparents, and provides a window into the times through which they lived. Here I explore the paths that led them to found what is now the world's largest and most successful animal welfare charity from a back room of their own home, with sparse funds, and at a time when caring for animals and our planet was seen as both crankish and sentimental. It is a story of rebellion against monocultures, against the factory farm, and against the reckless intensification of agriculture. It is also a story of personal and spiritual transformation.

I grew up just a ten-minute drive away from my grandparents' home, Copse House. It was CIWF's first headquarters, and as a family we spent great swathes of time together. Their home in many ways felt like my second home, and yet despite our close-knit relationship it turned out there was still so much for me to learn about these two extraordinary people. Though I was lucky enough to know them into adulthood, I had not, until researching for this book, fully understood the radical scope of their environmental, welfarist and compassionate vision. I had only known some of what they had achieved, and only a little of the extraordinary impact they'd had.

While it has been a privilege to be the person chosen to write this story, it has also been a joy for me to delve more deeply into their lives, and those of their peers and fellow activists. It has been an emotional, gratifying, and enlivening journey to understand their legacy more fully throughout the winding journey of this creative process, and I hope that you too will be inspired by their lives.

Emma Silverthorn

Foreword

Hampshire, England, in the mid-fifties, and two people's lives were about to change for ever.

One of them was a farmer, out tending his cattle in the summer sun, his hand resting gently on their black and white backs as he readied them for the day's milking. Their tails swished and flicked. They gave him a passing glance before bowing their heads for one more mouthful of grass before heading off to the milking parlour. For Peter Roberts, the normality of his farming day belied the events to come.

When the long day was done, he donned a tweed jacket and made his way to the neighbouring pub of white walls and rippling terracotta roof set back from the roadside. Outside the pub, an empty frame stood where a sign should have been. The sign itself had long since gone, blown away or stolen so many times that the landlord had given up replacing it. Hence this farmers' local became known as the Pub with No Name.

That summer, a family of townies from Tolworth, Surrey, pitched up beside the inn for a caravan holiday. They were attracted by the riding stables nearby, keen to get some practice in ahead of their next show jumping in Cobham.

That was how Peter met Anna, a pale, petite young woman with brown hair and an infectious smile. It was a whirlwind romance. Engaged within a week, they were wed two months' later in a marriage that would last more than half a century. That summer's day in 1955 altered the course of their lives.

During their farming years together, Peter and Anna continued to frequent their local. Tall, tanned and rugged, Peter had become used to shuffling below the low ceilings and dodging the beams. The couple passed the time chatting with fellow locals over a pint or two. This was the place where local farmhands and gamekeepers gathered to swap stories of wildlife they'd killed. But Peter was different. He farmed with innate respect for other living beings. He saw the countryside as an ecosystem of fellow creatures. He saw farming as part of nature. He could see how the dots joined. He had a strong sense of empathy and would later come to embody the word 'compassion'.

Though they both loved the country life, their days in farming were numbered. The countryside was changing around them. Farming was going in a new direction – and in Peter's view, not for the better. Those changes would alter the course of his own life, leading him to turn away from being a practising farmer, instead becoming one of 'modern' farming's staunchest critics.

As Peter and Anna were first meeting over his Hampshire farm gate, so a whole suite of influences were ushering farming toward intensification, something that many found themselves swept along by. Rural communities were close-knit and keen not to get left behind. There was a tendency to group-think, which helped embed new changes surprisingly quickly in the countryside.

Post-war Britain was understandably eager to shrug off wartime rationing and shortages of food. A new way of farming was being modelled in America. US aid for rebuilding war-torn Europe enabled countries to buy American food, then the means to produce it, using new intensive methods. Artificial fertilisers and tractors were shipped across the Atlantic. In Parliament, the 1947 Agriculture Act enshrined farming's new direction of intensification. Age-old crop rotations, natural pesticide control and animals roaming the countryside as an implicit part of sustainable farming came to be seen as old hat. They were replaced by abundant supplies of fertilisers and pesticides churned out by chemical factories. Farmed animals and their manure – long seen as an integral part of keeping soils healthy for future harvests – began to disappear from the land into cages and crates. They were turned into machines. Farming started to scale up and specialise. Agricultural subsidies helped fund this new 'modern' way. Agricultural colleges taught it to a new generation of farmers. Sales reps got busy selling the means to the new way – artificial fertilisers, chemical pesticides, bigger tractors, cages and crates. MAFF – the Ministry of Agriculture, Fisheries and Food (now Defra) – led the charge, with advisors fanning out across the countryside to get everyone on board.

In the sixties, the man from the ministry knocked at Peter's door. They had a long conversation, but the message was simple: if you want to boost your business, you'll have to move into rearing chickens intensively. Peter was told that meant specialising in chickens, lots of them, in large industrial sheds. He'd have to build a factory farm. He could buy the birds and their feed from a big company, and when they were fully grown – which would not take long – he could sell them back to the same firm, who would have them slaughtered and find them a market. It would be sanitised, industrialised, integrated. All he'd have to do was sign a contract and grow the crop of chickens.

Although he kept a few hundred chickens already, Peter felt uncomfortable. He felt this new way would mean relinquishing his power as a farmer to decide how things were done. It didn't feel right. He worried, too, about what it would mean for the chickens. His were free-range.

That evening, he discussed it with Anna. They made a good team; his careful, considered way of weighing things up were complemented by Anna's more intuitive approach. Her reaction was characteristically instant: 'If you want to do this, Peter, I won't stop you – but I want you to know that I don't agree with it.' The overtures from the man from the ministry were turned down.

Unlike Peter and Anna however, many others succumbed to the sales patter, only to find themselves on the treadmill of intensive farming that would drive many out of business.

That was the moment the Roberts family decided to leave farming and move to a country cottage in Greatham. Never wanting to be far from green fields and open countryside, theirs was the last house on the left before the village petered out into farmland and forest. In their new home, they would raise a family of three girls, Judy, Gill and Helen. It was there that they combined child-rearing with founding an organisation that would take over the rest of their lives: Compassion in World Farming.

And that was how a farming family transformed themselves into a new breed of countryside evangelists, warning others of the unintended consequences of factory farming. At the time they became outcasts, seen by fellow farmers as turncoats. Feelings ran high – so much so that when, even fifty years later, I found myself giving an address to Peter's former local farming club, the pent-up resentment and anger from the audience was palpable. I could feel unspoken in the chairman's vote of thanks the words 'You're lucky to get out alive!'

Back in Peter's farming days the world had seemed very different. First, it had a population of just 3.3 billion people.[1] Climate change wasn't really talked about. The number of farmed animals produced for food each year globally stood at 10 billion, with meat being eaten much less often than it is now; when my mum started buying chicken for Sunday lunch in the early seventies, it was seen as an aspirational treat. Contrast this with 2020, by which time the world's population had reached 7.8 billion on a planet on track for a temperature rise of

1 Worldometer, *World Population by Year* (worldometers.info website), https://www.worldometers.info/world-population/world-population-by-year/

around 3 degrees Celsius.[2] Despite animal farming being responsible for more greenhouse gases than all the direct emissions of the world's planes, trains and cars put together, our appetite for meat has grown to the point where 80 billion animals are reared and slaughtered for food each year, the majority in factory farms.

Even back in the fifties Peter had seen the way things were headed: more and more animals in confinement; much of the world's croplands diverted to feed them. The feedstuffs themselves were increasingly being produced in ways that drove out and killed wildlife, and damaged the soil – the very foundation of farming in the future. He was alive to the false rhetoric of farming's new way, that it was about 'feeding the world' when huge swathes of cropland that could be feeding people were instead stoking factory farms. In an address in the seventies, he spoke of how there was no actual 'shortage of food in the world', but rather that there was starvation 'because we devote the major part of the world's resources and its expertise ... to the feeding of animals instead of children'.

By the eighties, during Bob Geldof's monumental Live Aid concert in aid of famine relief in Ethiopia, Peter was pointing out that vast amounts of cereals were being *exported* from Ethiopia to feed intensively reared animals in other countries. He was saying the things no one wanted to hear; for instance that '370 million tons of the world's harvest is fed to livestock – enough to meet the total combined needs of the population of China and India'. He went on to predict that if we continued along these lines, 'famine would increase on a scale never before seen' and that this would be followed by 'the collapse of order and finally in war'.

Peter was ahead of his time. He identified the connections between food, farming, famine and the soil. I remember one of his seminal leaflets, 'Aims and Ambitions', in which he described why it was important to move away from diets heavy in animal products and particularly those from grain-guzzling chickens and pigs, which consume food that would better be fed directly to people: 'We must progressively reduce our dependence on animals and gradually relegate them to the less productive land. As we do so we must shift the emphasis away from monogastric animals.'

I always felt that Peter should write a book spelling out the connections between farm animals and nature, the soil and us, but he never did. So, thirty years on, as his protégé and inspired by his extraordinary vision, I found myself writing books with titles straight out of his predictions: *Farmageddon*, *Dead Zone*, and *Sixty Harvests Left*.

2 *Global Biodiversity Outlook 5* (Secretariat of the Convention on Biological Diversity: Montreal, 2020)

Today there is much talk of impending climate and nature emergencies exacerbated not least by the industrialisation of food production and the threat of new pandemics arising from the way animals are treated. The UK government is recognising animals as sentient beings in law and thinking through ways of redirecting agricultural subsidies to encourage more nature-friendly farming. The EU has announced it will ban all cages for animals farmed for food. And the UN Secretary General has called a global summit to talk about an urgent 'transformation' of food systems globally.

At last, a lifetime since Peter and Anna first met, the penny seems to have dropped: *the way we farm today has a profound impact on all our tomorrows.*

In this book, we go deep into the heart of that family to find out what motivated them, what inspired them to stand out from the crowd. It leads us through their lives, their trials, their tribulations and their sacrifices. It unearths their personal and public battles, conducted with a fearless integrity, not least of which was in taking a monastery to court for keeping calves in veal crates, a battle they lost only to win the war; the UK government subsequently made veal crates illegal.

In this book, we discover what it was like to be an early pioneer advocating concern for all creatures, whether human or non-human. We find out what it was like to endure anger and ridicule, to be labelled 'cranks'. This is a particularly human story of ordinary people that did extraordinary things. Above all, this is a tale of a couple's undying love for each other, and their compassion for all, encompassed in a vision based on forecasting farmageddon.

Philip Lymbery
CEO, Compassion in World Farming

Acknowledgements

Thanks are due to CiWF and its trustees for supporting this book in its germination and to the archive department at CiWF.

A heartfelt thanks goes to all of my family, near and far, for giving me to access their memories, recollections and photo albums and to all those colleagues, activists and friends of Anna and Peter who allowed me to mine them for stories. Thank you to Peter Egan, Sue Jameson, Tess Burrows, Kim Stallwood and Emma Milne for their kind endorsements.

I am grateful to those who have helped me with the marketing side of things, it is much appreciated. Thanks also to my agent Robin Jones and to Whittles Publishing for taking a punt on me! Caroline Petherick is due much praise for the excellent copy-editing.

Thank you to all family and friends who continued to support and encourage me even though this took much longer than expected - you know who you are - and to my adopted alley cat Stevie for everything you naturally do and bring. Above all thank you David for your wisdom, love, support and clarity upon which I have relied so much.

1

THEY JUST KNEW

Edna Dorothy Hearsey was much more interested in horses than men – or so she'd thought until she met a certain farmer on a family holiday that summer …

The Hearsey holiday caravan had been stationed in the field of the White Horse pub next to a farm in rural Hampshire. I imagine them both as they were back then, decades before I, one of their six grandchildren, was born: my grandfather, Peter Holtom Roberts, packing his pipe with a large pinch of tobacco, and at the local stables my grandmother, Edna, mounting a plump-bellied horse, which takes her out at a canter across the fields. Peter, leaning against the stone wall, spies her; there is no one else in sight except a couple of circling magpies. He sees she is pale, petite and upright, and that a few strands of brown hair curl from under her riding hat. He strikes a match, only looking away from her twice: once to char, once to re-light, his pipe.

That evening the Hearsey family walk the few yards from their caravan to the White Horse for drinks and food. As they enter, they pass a tall pole with an empty frame swinging in the breeze. The landlord tells them the sign has blown off or been stolen so regularly that landlords past and present have given up replacing it, hence its unofficial name of the No Name. Softly lit, with low ceilings, the room has exposed brickwork, knotty beams and candle-blackened walls.

Edna reads a framed poem on the wall, 'Up in The Wind' by local poet Edward Thomas; something about the wild spirit of the land and the yodelling of curlews. The latch of the creaking door pulls her away from the verse. She sees him come in, broad in his checked shirt, a few buttons open at the top, his face rugged with large features that remind her of Welsh actor Richard Burton.

'Good evenin'. Here on holidays?' He raises one eyebrow, removing his pipe to speak, then putting it back in to his mouth to shake hands, his tanned and warm, twice the size of her small, cold-fingered ones. Her eyes are a light blue; she sees that his are inky brown.

'May I?' he asks in his well-spoken Attenborough-ish tones.

He pulls up a stool, and a chocolate labrador who he introduces as Meg settles cosily at his side. His voice is clear, deep and proper: hers more suburban. He talks about life at his farm, his father the doctor, and his violin-playing, Shakespeare-loving mother. Edna is at the periphery of his vision as he turns to her father Fred, and she is the driving force behind his words when he addresses her mother Dorothy and her younger siblings Jan and Freddie.

He learns that the Hearseys live eleven miles from Charing Cross, in Tolworth, Surrey, and that Edna works in stables and as an office clerk. She adores horses. Her father gardens for a living, her mother cleans. They get one week holiday per year, and this one is doubling up as a house-hunt; Fred grew up in Hampshire and for years has yearned to return.

On the trundle back from the White Horse to their caravan the dewy night grass wets their shoes right through to their toes. Edna struggles to sleep, her family surrounding her, the handsome farmer just a field away. Back at the No Name, the velvet burn of last orders brandy warms and stings his chest as he too thinks of the person he's just met.

In the morning light she dresses quietly and goes out. She climbs over stiles and her feet sink into the clay-rich fields. Greens and chestnut brown are followed by flashes of purple, scarlet, and mustard. Lilac and buddleia spring from the hedges, attracting red admiral butterflies; wisteria climbs grey stone cottages, and the strong perfume of honey-mustard-musk oozes from the expanses of rapeseed fields.

Later, Jan joins Edna for an afternoon of fruit-picking. Soon the sisters' forearms are stained with juice and nicked with brambles. 'Take the berries to the farmer. He said he liked blackberries last night,' Edna instructs her sister. Jan is curious; her sister has never had much time for men. She much prefers horses and doesn't suffer fools gladly.

Jan smells the farm, sour and earthy, before seeing it. Inside the gates the cows in stalls are chewing, their udders attached to steel milkers, the whirring sound of a generator mixing with the odd moo. The farmer appears, rolling a large steel can across the yard.

Jan approaches him, blackberry punnet outstretched. 'Edna sent me,' she explains. Peter smiles, takes the punnet. Sweet and sharp, they are the best blackberries he's ever eaten.

A few days later he brings the family fresh eggs from his own chickens and the paper for her father, then invites the whole family over to his farmhouse, Little Barnet, for afternoon tea. They sit by an enormous open fire, lit only for the kettle; beyond the hearth the old farm cottage is stone-cool. The kindling crackles.

Peter asks her parents if they'd like another cup, but Edna is quick to butt in: 'Don't worry about *me*, then.'

'You shut up,' Peter responds.

'Please shut up,' she retorts.

This is flirting fifties style.

The pair stay up late, the rest of the family finally falling asleep, and Peter furthers his case with, 'I'm after a good housekeeper. Would you be keen on the position?'

'Oh, there's a vacancy, is there?' Edna replies.

He asks her again, properly this time. She gives an assured, instant yes to his proposal of marriage. They have known each other for but five days.

The next day she has to go back to Cobham; she's jumping a horse there. Her fiancé buys her a good luck neckerchief for the ride. In the photo of her by the horse she looks proud and a little shy, in a thick tweed jacket, black jodhpurs, a boyish black tie.

The horse is called Black Pirate and, true to name, throws her off at the first jump. The commentator blares over the tannoy, 'Black Pirate has made his rider walk the plank.'

'A good rider must be thrown off seven times and get back on … of course,' she recalls Peter telling her.

On the coach journey from Edna's family home in Tolworth to the market town of Petersfield, close by Peter's farm, she and Jan eat butter and jam sandwiches and drink strong tea laced with the plastic tang of Thermos. They arrive at Petersfield's market square, where a statue of a conquering man on a horse stands at its centre. Peter meets them and leads them straight to the jewellers, just off the High Street. They peer into the glass counter: row after row of gold, jewelled bands sit in soft velvety trays. The clerk removes a key from the jailer's set jangling from her waist and undoes the glass. Edna knows instantly. It's beautiful, with a large smoky sapphire at its centre and a small diamond on either side. He slides it onto her ring finger; her fingers are slender but have large, knobbly knuckles and short squat thumbs which she tries to hide from him in her palms.

Ring bought, they spend the rest of the trip walking the fields and woods that seem to fill almost the entire county. On reaching the centre of a four-acre field they've just crossed Peter wonders if they've latched the gate at the other side. Latching gates is the number one rule in the Country Code rulebook. Sister Jan is sent to check. The sun is warm against their backs. Alone, chaperone gone, they kiss.

On a chilly November day, under a bright blue sky, Edna, on her father's arm, walks up the flagstones of the Hearsey family's parish church. Traditionally attired in white, and veiled, she is a pretty cake-topper. Untraditionally she requests that the reverend remove the word 'obey' from the bridal vows. It is 1955, and the official beginnings of Second Wave Feminism are five years away. Edna Hearsey becomes Mrs Roberts.

Peter holds Edna by the waist as she steps into the grand wedding car taking them to the village hall for their reception. Fittingly for a farmer and his new wife, the hall is covered with children's drawings of cows and geese with the appropriate word for each animal below. Bowls of jelly and plates of starchy, pastry-wrapped food adorn the tables, and his hand covers hers as they cut a three-tiered thick-iced white cake. Then, after a week's honeymoon in North Wales they begin their life together at Little Barnet Farm.

This was the story told on a loop at every family gathering and celebration, the mythos of their great and instant and forever love shaming us all with our messy modern relationships, divorces, uncertainty. The numbers involved even had a lovely symmetry about them: five days after meeting they were engaged, ten weeks later they were married, and their marriage lasted for fifty golden years. Edna, my nan, (who later changed her name to Anna and who will be referred to as Anna from here on out), adored romance novels and no wonder; her own love story was as idyllic and as simple in plot as any. A summer holiday (as in the film *Grease*), a working-class girl (à la Catherine Cookson) who is fiery and unromantic (think Elizabeth Bennet in *Pride and Prejudice*), a rural setting (Austen and Brontë), meets a middle-class boy (moving up a station, that's Austen), who lives on a farm (*Darling Buds of May*) who sees girl on horseback (Jilly Cooper but minus the smut), and falls instantly in love (every single story). There is an exchange of wholesome edibles and afternoon tea (Austen again), with a romantic comedic ending (Shakespeare, Austen, all of them; marriage is the law of the genre). Wagner's Bridal March plays, and off they go into the distance (it's the Disneyfied version of a fairy tale). It is only death which parts them, a week before their fifty-first wedding anniversary – but then again, they believe in the afterlife and so are never in their views never really parted (this part Gothic and mystical as in *Wuthering Heights*, but without the revenge bit).

I hold them up as a beacon; tell their tale to friends, who are duly awed and perplexed. Some want reasons and logic. That was the era of fast marriages, they suggest. Hastiness was more common in those days, they continue. But a five-day proposal has never been that common, and quick marriage was a pre-war trend, not strong in the post-war fifties, which was when they met.

Others cite youth as the reason, with all its connotations of impulsivity and romantic giddiness. But Edna turned twenty-nine the day after their engagement, and Peter was thirty-one during their brief courtship. Religious duty? No, neither had much of that, at that point anyhow, only following the Church of England in the most cursory of ways.

Their own answer, when asked, was this: it was love at first sight, the thunderbolt; without a doubt 'they just knew'.

So begins the biography of Anna and Peter Roberts, the founders of British charity Compassion in World Farming, today one of the largest international animal charities in the world, which still fights to end what is by far the biggest cause of cruelty to animals on the planet.

2

BEFORE THE NO NAME

Before Anna and Peter met that night they had each lived three decades of life separately. Peter had been born in 1924 and Anna in 1926. Each had lived through a war, and found what they believed would be their life's calling. Here there was already some alignment, for both were based on the land and animals: for Peter it was farming, for Edna, riding.

Tweed-jacketed and sharp-featured, Captain Leslie Douglas Roberts, pipe clenched between his teeth, had made it clear to his two sons, Peter and Frank, that they would grow up to be doctors. His own baptism of fire – first year service with the Army Medical Corps during the Great War, followed by a respectable career as a country GP in the Midlands – made things, to his mind, perfectly simple.

For a while Peter did try to follow suit. At age eleven he was sent off to boarding school, a good school in terms of league tables, that his father no doubt believed would set him on the right track for medical training. I picture my grandfather as boy shortly before leaving home for the first time. I see him, swinging a stick left to right, occasionally slashing a fern on the side of the dirt track, eleven years old enjoying the cool of his mother, Emmie's, hand on the back of his neck. She explains again about Denstone School in Uttoxeter, where he'll be going in a fortnight's time. How this would be the last of their summer holidays spent wandering in the forest at Cannock Chase, close to their home in Rugeley, Staffordshire.

'You'll stay in a big room with the other boys your age – but that'll be fun. You'll make lots of friends,' I hear her say. All the stories of Peter's mother emphasised her sweetness, kindness and intelligence.

Peter, in going off to board at Denstone, was following in the footsteps of his elder brother Frank, and was leaving behind his elder sisters, Mary and Barbara, as well as his parents.

The school grounds contained acres and acres of green, and the school itself was Victorian and very grand, just a half hour's drive away from the family home.

An old report card of Peter's shows that he did well in physics, literature, mathematics and scripture; he was intelligent and had a love of learning, and as a child would learn lines of poetry by heart; some of his favourite authors included William Blake and John Clare. As well as this, he trained himself to recognise particular bird songs from a young age.

In the school holidays young Peter made the choice to go and work on local farms, a decision that would impact his life more greatly than he could have imagined as a teenager. Here he enjoyed something different from academia, the rough wood of the hoe in his hands, the heat and ache in his arms and back a relief from the scholarly pressures of Denstone and his somewhat overbearing father; he was barely getting paid, but it didn't matter. There was turning the soil, that primordial scent, and later seeing the first green shoots pushing up into the light, spreading handfuls of soft hay and mucking out the animals. At that time he began to imagine a farm of his own one day, not here in the Midland flats, but high up on the Welsh hills looking after sheep on 1,000 acres, roaming wild and free. But then the captain's, his father's voice came back, Pops, as his children called him, had made it clear that he and his brother Fred were to be doctors and that was that.

Peter was fifteen and still at Denstone in 1939 when World War II broke out; when it ended he was twenty-one. Two years after the war's official end, in 1947, Peter completed some years of military service, the details are somewhat unclear, particularly as, like many men of that era, he rarely spoke of it. However, what he did mention on occasion was being sent to Malaysia to aid in the transportation of a pack of mules from East to West. In a letter he stated that the mules were being sent to 'help fight the Communists in Greece'. He was an officer, in charge of men and mules on this arduous and lengthy mission. In the monochrome photographs of this trip the mules can be see bucking all the way down the boat-ramp at the port of Rangoon, in Burma, the whites of their eyes flashing and the muleteer only just keeping hold of the reins, while clods of dirty engine smoke are puffing into the sky above the ship.

Mules of course aren't meant for sea travel but they are notoriously tough animals and at this point would have already survived the Hobday procedure (in which the creatures' vocal cords are cut in order to end all that loud braying, so that they can be deployed during times of conflict), as well as of course the war itself. The Hobday procedure is described in archaic veterinary manuals as 'a simple matter' in which the mule is injected with a sedative, following which their vocal cords would be seized between two pairs of forceps in order to make a V through which a pair of blunt-nosed scissors would cut, leaving a raw strip across each side of the larynx. Following this muting of the mules, the animals would have been chucked out of a Dakota aircraft and into the jungle (with a parachute, though many still broke their legs on the way down), and then, if they survived all that, into as many years of conflict as possible. The mules Peter helped transport may well have been some of famous General Wingate's mules, the general whose guerrilla tactics had been so vital and pioneering during World War II. During the war mules had been employed internationally from America to India, Japan to the Mediterranean, and now it was Peter's duty to take them through the Suez Canal.

Peter, who already had an interest in spirituality, bought a book in Burma called *The Light of Asia* by Victorian writer Sir Edwin Arnold; this text, as it turned out, had nestled amongst its lines of verse views on the eating of animals – views which were to greatly impact Peter's choices later in life. Peter and his friends and fellow soldiers also visited many Buddhist temples. His photo albums are full of images of great reclining golden Buddhas and fierce *chinthes* (leonine figures that guard the Buddha). Incense would have wafted around the statue's huge clawed feet, and saffron-robed monks are pictured floating to and fro, their hands full of rice-ball offerings, a few feral dogs stretched out enjoying the cool of the temple floor. Rangoon temples were undeniably a world away from the religion of Denstone's chilly Gothic chapel.

Presumably en route to Greece, the ship stopped for several days in Aden. In the photos a volcanic mass lies jagged between the dark of the gulf and the cloudless sky, and further into the album are photographs of what looks like military training exercises and a mission involving taking equipment around a colossal crater and high up into the mountains.

Back on the ship Peter noted how the mules aboard were 'treated like VIPs' and in one picture a mule is shown wearing a lady's floral sunhat, with two holes cut out for his ears, both protecting him from the blazing heat and providing entertainment for the men.

Although Peter rarely discussed this wartime experience, transporting that pack of mules and his travels in Burma, Yemen and Greece were undeniably a formative experience. When he wrote a letter much later to his grand-daughter Amy for her primary school World War II project, he described to her the painful Hobday procedure that these animals had had to undergo (albeit in child-friendly terms) and noted also how he had 'gained quite a good rapport with those mules' over the course of the long journey. Tellingly, he finished his letter explaining how 'he'd dreaded to think' what would happen to the animals after he'd let them out of his care, and how he was worried, as mules 'were looked on as very inferior animals by the Greeks'.

Something about that heightened post-war period seemed to increase both his desire to farm animals yet also, somewhat ironically, increase his empathy for them. Before pursuing his dream, however, he decided to make some final attempt at fulfilling his father's wishes, and he gained a place at King's College Medical School in London.

There are but few tales to tell of Peter's time there, as he quit after one year. The overall feeling was that he had never had any desire to be a doctor in the first place, that it was largely paternal pressure that had led him to sign up, and that from his teenage years onwards he loved being on the land above all else. The one story I recall regarding his brief sojourn in London was how one night after drinking a lot of beer with some friends he had mounted and ridden the famous bronze lions of Trafalgar Square. When I moved to the city in my early twenties he advised me to give it a go! Lions aside, after that year he did find the courage to go against his father's wishes, and switched track to the one he had always known he ought to be on. He signed up for agricultural college.

Opened by a 'gentleman farmer' at the tail end of the 19th century, Harper Adams College boasts 178 acres of farmland within its campus, and is one of the country's most prestigious agricultural colleges. Peter managed to obtain an officer's grant to attend Harper Adams, and reluctantly his father came to accept that neither Peter nor Frank would follow in his footsteps; Frank studied engineering and later worked in India and Hong Kong as an engineer.

Peter seemed to do well at Harper Adams, gaining valuable academic as well as practical hands-on knowledge of farming, though he would later claim that during his three years of study his main distinction was becoming captain of the darts team. In a black-and-white photo Peter sits on the grass at the front of the grand Victorian building with three friends, smiling in the sun, Peter is cross-legged in a lotus-style yogi pose, his fingers resting on his knees in the Om shape, a less than holy cigarette hanging loosely from his mouth. He had had an early interest in yoga, solidified by his time in Asia; in fifties rural England, it's worth noting, yoga was seen as a rare and esoteric practice. After one year of academic and two years of practical study at Harper Adams, Peter graduated, enthusiastic to begin his life as a qualified farmer.

Anna as a child had had a similar interest in nature and a similar love of literature – but that is where the similarity in Peter and Anna's childhoods seems to end. While Peter's was decidedly middle-class – private school, educated into his twenties, and with a doctor for a father – Anna's was working-class. But like Peter she loved to spend as much time as possible outdoors, pottering in the garden, though she was not allowed to get involved with actual gardening; her father was a gardener, so that was his job, after all, and he was very particular about it. She might be allowed to water the flowers of an evening but other than this privilege the garden was out of bounds to everyone but Fred. Her mother cleaned professionally, and worked hard at home too, as a wife and mother. Their home was modest and seemed especially small when compared to their Aunt Maud's home in Hampshire. Her father's sister, Maud, had a large country house, The White House, the result of a financially prosperous marriage; it had no other houses anywhere near it, a novelty for Anna whose family home was terraced.

The seeds of Anna's love for the countryside were planted at Maud's. Early adolescent Anna had been evacuated with her brother Freddie, which means being sent there without knowing how long they'd be gone from home, and told they'd stay there simply 'until it was safe again'. In Tolworth the effects of the Blitz weren't as bad as in the centre of London, but there was still a great deal of danger living in Greater London especially near the end of the war when the sweetly named but terrible bombs known as doodlebugs had been dropped on the town several times.

Despite the terrible reason for being sent away Anna had many wonderful memories of that time: the lush garden, the bucolic landscape and scrumping apples from a local farm … The downside of this stay, beyond missing her parents, was that Maud was married to a man who Anna's family tended to describe as 'miserable' and 'ancient'. His name was Alfred, and he did not like children; he was by all accounts an anachronistic man who believed that children should be seen and not heard.

From a young age Anna found solace in literature; books kept her company and offered her a retreat. At home she loved to read, and at night her father would bellow up to her bedroom 'Lights out!' to which she'd reply 'Just *one* more chapter … ' Thirty minutes later, the light still on, he'd call up again 'Bloody long chapter.'

Yet, despite Anna's love of learning and particularly her love of English literature she had to leave school at fourteen, as normal for a working-class girl of that era. Anna's life as an evacuee, however, came to an abrupt end, totally unrelated to the war itself, one sunny day, when she was caught chatting to some young boys at that local farm, the Wangaree Fruit Farm, an act which apparently led to her being chastised by either her mother or aunt or both and which resulted in her being sent back to Tolworth. (This event she saw as incredibly unjust and unfair, to the extent that in her late seventies, on more than one occasion and after a couple of very small glasses of white wine, she would still complain about it to me and her second daughter, Gillian, my mother.)

But Anna would be returning to Hampshire. For Anna's father, after all, it was a dream place and he would repeat, mantra-like, his aim to 'get back to Hampshire given the chance' one day. 'The Hearseys,' her father went on, 'dated back a long way in Hampshire.' He was right; in the village of Greatham, where Anna would later live with Peter, there were crumbling Hearsey graves dating back over 200 years.

Once she was old enough to join in the war effort, Anna signed up for the Ladies Army. This meant swapping her black jodhpurs for a thick khaki woollen dress, and her velvet riding hat for a Victory Roll. In a training camp in Guildford, she learned how to dress wounds and give injections, and learned about the signs of dehydration and gangrene. She continued her training down in Devon, where as well as learning the basics of medical care, there was a chance to walk on sandy beaches, one hand holding an ice-cream cone, the other shielding it from the gulls.

She was eventually stationed in the north-west London suburb of Harrow, where a bomb dropped on the camp. She never talked about this at any length, only saying once or twice that it was 'absolutely awful'. After this she was posted to London as a nurse, where she served one year. She was glad to leave, not only owing to the visceral and gruelling nature of war service, but also owing to her strong dislike of busy, nature-deprived, city life.

Then at midnight on 14/15 August 1945, Prime Minister Clement Atlee made a radio address to the nation, announcing that the war was finally over. 'The last of our enemies is laid low,' he stated confidently. On the evening of the 15th, VJ Day, King George VI broadcast to the nation at 9 pm from his study at the Palace, saying, 'Our hearts are full to overflowing, as are your own. Yet there is not one of us who has experienced this terrible war who does not realise that we shall feel its inevitable consequences long after we have all forgotten our rejoicings today.'

The celebrations were even more exuberant than on VE Day back in May, and Anna, still in the capital, joined in outside Buckingham Palace.

3

WONDER CHEMICALS AND CAGES

The war over, Peter and Anna planned to continue as they had been pre-war, he with his plans for a rural farmer's life and she riding with Jan as much as she could in her free time, while working as a clerk during the week. Peter began to search for a farm of his own per his adolescent plan, though instead of remote Wales he landed on a suitable one in south-east England. While both settled down to fairly humdrum civilian lives, the British landscape was undergoing a massive and radical government-backed transformation – a transformation that would, it turned out, intersect deeply with Anna and Peter's own life stories. After years of rationing, parts of it remaining in place until 1953, the government had instituted extensive measures to maximise food production, the aim being to increase agricultural output by a massive 60 per cent compared to pre-war levels.

The best way to do this, the government's various bodies decided, was to mechanise the countryside. Space, fresh air, natural diets and sunshine were suddenly deemed unnecessary luxuries for farmed animals. They were taken off the land and put into cages and crates indoors, either in darkness or under artificial strip-lights and fed artificial, highly processed diets, sometimes starved, sometimes overfed, depending on production needs. Under the new systems laying hens were put into squalid battery cages, broiler chickens (those used for meat) into cramped, humid, high-density enclosures, pigs into narrow (sow) crates, or barren, overcrowded pens; dairy cows were taken off the pasture and kept indoors for much of the year, made to produce unnatural and painfully high yields of milk, the surplus calves required for this yield – the males, that is – imprisoned for their entire lives within narrow, barred veal crates. More and more animals were selectively bred, something farmers had done for millennia, but under this new high-pressure production system the practice developed monstrously and began to reach untenably cruel levels.

Breeding choices made to produce better meat animals often lead to the extreme detriment of the animal's health – and often, as a consequence of that, to the detriment of the humans that regularly consume such sickly animals. The broiler chicken is one of the most extreme examples of the perils of extreme selective breeding; with their oversized breasts most meat-chickens either suffer pain whilst standing for even short periods of time, or have severe heart problems, or are so lame that they cannot stand at all.

The process of industrialising farms that began during this post-war era has continued and accelerated to this day, in the early 2020s. A post-war moral gulf had opened up, and many agricultural institutions and Western governments now appeared to see no difference, as the *Daily Mail* newspaper put it in 1960 between 'mechanising a plough and mechanising an animal'.[3]

The post-war shift radically altered farming practices, and indeed the lifestyles of many farmers, moving the profession – or as Peter had found it, 'the calling' – from the noble pursuit of working the land and feeding hungry people to, instead, one of producing more and more food regardless of its quality, regardless of whether or not it was to be eaten. This hyper-production is to the detriment of so much, especially our environment, animal welfare, and farm workers' health.

The aim had been an understandable one at first, aiming to fill a gap after the meagre years of war, but hyper-capitalist tendencies ended up overriding this reasonable aim. Individuals and families' diets began to be affected and influenced by the politics and businesses behind this post-war industrialisation, and meat moved from being a once-a-week treat – 'a joint on a Sunday', as Peter put it – and maybe a small piece here and there during the week, to an every-meal apparent necessity. Animal flesh changed from luxury to expectation. In an early Compassion in World Farming newsletter Peter wrote how 'The industrial approach to agriculture since World War II had led to widespread abuse of animals and the land; the system was broken and dangerously expansive, as the mass production of animals' meant the need for 'a plethora of ancillary industries to supply buildings, equipment, concentrated feed and drugs'.

Peter described this post-war shift from the aim of achieving food security, succinctly summing up modern agriculture's ethos as: 'If it moves, kill it; if it doesn't move, spray it.' [4] He went on to note how we had landed in a position where now 'more than half of all the world's food either rots, gets dumped

3 *Daily Mail*, 16/7/60.

4 Peter Roberts, *Agscene* magazine, 1968, Winter Issue, p.3.

in landfill or feeds imprisoned animals' – a position where greed too often provides motive beyond all else. Peter was an early critic of this fairly new system, pointing out its lack of sustainability and its gross inefficiencies, as well as the damage it caused and would continue to cause in terms of pollution, disease, and even famine.

During this time, the land itself, as well as the animals that had once grazed on it, underwent further industrialisation; monoculture began to replace mixed; rotational farms and forest and wild lands were seized for more and more animal agriculture and feed. Synthetic chemicals began to be relied on much more heavily than ever before, in both animal and arable food chains. The soil was saturated in DDT, an apparent 'wonder chemical' developed as a weapon during World War II. As environmentalist and activist Rachel Carson put it in her seminal book *Silent Spring*, published in 1962 and first serialised in *The New Yorker*, the idea that DDT could be used as an insecticide, (or, as she says more accurately, a 'biocide' – a poisonous substance that causes the destruction of life) did not come by chance; insects in labs had already been 'widely used to test these chemicals as agents of death for man'[5]

Carson was an American marine biologist, writer and conservationist whose influence is still being felt today. In *Silent Spring* she criticised the profligate use of DDT and other dangerous chemicals which was now soaking vast quantities of the countryside of the Western world, linking their use to a devastating loss of wildlife; she also demonstrated how these chemicals would enter the human food chain, and the untold genetic damage they could cause. Carson particularly emphasised the threat to the ground beneath our feet that these chemicals posed. The nucleus of all life, as she expressed it, was the soil, it was the natural world and humanity's life force. She asserted what perhaps now seems obvious (to some of us, at least), that the soil is supremely valuable, beyond economics, that it is linked to everything else in the food chain, that it is the ecological equivalent of breath in the body. Put simply: if there's a deficiency in the soil then all else suffers. Over the decades since its publication *Silent Spring* has proved not to be a case of environmental paranoia or hysteria, as some of her critics suggested at the time, but a prophecy coming true daily before our eyes.

Despite some detractors Carson's theory became widely known amongst the counter-culture and environmentalists of the era – so much so that by 1970 folk singer Joni Mitchell was singing to the farmer 'to put away the DDT now' in her hit song 'Yellow Taxi Cab'. As Carson predicted, and as good science showed, DDT did turn out to be so environmentally and biologically devastating that it was banned in America in 1972, in the UK in the eighties,

5 Rachel Carson, *Silent Spring* (Penguin Books, London, 1999), p.31.

and worldwide in 2004. Mountains of evidence point to DDT's carcinogenic effects on humans as well as its deadly effects on wildlife; worryingly, in the UK almost half a century on from the ban traces remain present in crops as well as in human cells. Despite the avid support of some at the time, Carson was threatened with lawsuits by chemical companies, accused of 'emotionalism' and 'gross distortion', and was even falsely accused of being a communist. This is less surprising when you know that in the USA Senator McCarthy was in power until 1957, and the pervasiveness of the so-called Red Scare remained into the early sixties. Carson, who died prematurely in 1964 and sadly did not live to see the positive impact generated by her work, has in recent years been experiencing a popularity surge, perhaps owing to the now frightening proximity of her early warnings and predictions.

Like Carson Peter understood, on a very direct level, the importance of what's beneath our feet. His time at Harper Adams had educated him about the properties of soil, and his private studies had led him to Carson's seminal book with its view of the soil as an incredible living organism. He had worked with the soil as a farmer and continued to do so once he officially quit animal farming in the sixties. His official job title in the late sixties and early seventies was 'soil chemist', and throughout this time he worked as a soil tester and expert for a Hampshire-based lime quarry business. In the school holidays he would take daughters Judy and Gillian out on the job, trekking across fields with test tubes, chemicals and packed lunches. The children were given the job of holding the tubes while Peter analysed the soil samples for mineral content and more.

He appreciated the interconnectedness of the soil's health to plant life, wildlife and human life, and was opposed to monocultures and the increasingly heavy reliance on synthetic fertilisers and insect deterrents. He stated on more than one occasion that if farming carried along the route it was then headed then in the future 'plant disease will prove more powerful than the bomb'. We are today in a position where many scientists, environmentalists and even the United Nations have predicted that if we continue using the land as we presently do then will have only around sixty good harvests left, this owing to soil erosion, degradation of fertile top soil and desertification. King George VI's sentiment on VE Day that, 'we shall feel World War II's inevitable consequences long after we have all forgotten our rejoicings today' has proved true in multifarious, unexpected and dreadful ways.

But all these prophecies and critiques of Peter's were to come later; at that time, in the mid-to-late fifties, Peter and Anna were all set to live an ordinary life as farmer and wife; the path seemed to be laid out: a simple rural, pleasantly quiet, life.

Anna would have been expecting this house and its attendant farm to be her home for years to come. This prospect suited her well; she loved the countryside and she loved Peter. Life was simple and happy but not necessarily easy. Water was still limited, the village only obtaining mains supply instead of a well a decade before, and there was no still no mains electricity. Most days she filled the kerosene Tilley lamps. The milking was done with the help of a large generator, and she always seemed to have the scent of diesel on her fingers. While the post-war boom of convenience produce took over urban and suburban areas, Little Barnet Farm was years behind in terms of domestic ease. Each morning she built the fire, then banked it at night so that getting out of bed was less of a shock at dawn. The whole of Monday was set aside for washday. It was slow living long before the concept.

On a cold late January night two years into their marriage Anna gave birth in the bedroom, a small room with uneven floors and views across the rolling South Downs. Judy Holtom Roberts had big blue eyes, Anna's brown wavy hair and pale skin. She had long legs and a short body. Later her party trick was being able to insert a whole Malteser into each of her nostrils: her nose was not especially large but her nostrils were Malteser-sized. Judy inherited Anna's unique (stubby) thumbs too. At a very young age she had witnessed a calf being born as she was being held by her Aunty Jan, Anna's little sister. 'Calve-y!' she had exclaimed. It was one of her earliest words.

Three years later as the leaves were turning, baby Gillian, my mother, was born, also in the wobbly-floored room. Gillian had small cat-eyes, a blend of Peter's brown and Anna's blue, and caramel blonde-brown hair. She inherited a strong nose from Peter, was petite like Anna, and also got the Hearsey thumbs. Later she would put her thumb in a matchbox, cover the bottom with ketchup and scare friends, claiming her fat thumb was a severed, bloody toe she'd found on the floor.

There wasn't a huge amount of leisure time at Little Barnet but when there was the family took trips out in 'JOG' – Peter's jeep. Peter seemed to have distracted Anna from horses as after they were married she rarely went riding.

The rest of the Hearsey family moved down to Hampshire, as planned, close to the Roberts. Peter's family were, however, more disparate, living across the Midlands then New Zealand and for a time Hong Kong. The Roberts had roasts together on Sundays and occasional trips to the pub. As children weren't allowed inside country pubs back then, the girls would sit in the car waiting for their parents to bring out tomato juice and crisps.

Whenever they left doors open between rooms, making the house chilly, Peter admonished them with, 'Were you born in a *barn*?' to which they'd reply in the affirmative. From Austen and Cookson the narrative had moved to the self-sufficiency exemplified in the BBC's classic comedy series 'The Good Life.'

And then the sixties came, and changed everything.

'It truly felt like Year One,' said English writer and radical Angela Carter of that tumultuous decade. Ethical shifts were complex but there was at least a boiled-down potted tale to unravel Peter's route from animal farmer to vegetarian animal welfarist. What occurred for Peter and Anna in the first years of that decade was what philosopher and animal activist Tom Regan had called a Damascene moment. He had believed there to be three basic personality types of animal welfarists: the 'Da Vincians,' who he said feel intuitively for animals and are 'born with an ability to enter into the mystery of the interior lives of animals'; Damascans, who 'undergo a dramatic and often instantaneous perception change in their attitude toward animals'; and Muddlers, 'who grow into animal consciousness step by step, little by little.' At different times, he had noted, we might encompass a combination of two or all three of these personae.[6]

Anna and Peter's Damascene moment came at Little Barnet. Like their great romance it is a story every member of our family knows by rote, by heart. One day a man from the ministry informed Peter that he ought to put his hens into intensive rearing, that it made good economic sense. In fact he said it was not only sensible but in fact vital. It was either expansion, which the Roberts couldn't afford, or accept the broiler, or go bust. The man offered his apple, like the wicked witch in Snow White.

Later that night Peter relayed to Anna what the man had told him.

Her response was as quick and sure as it had been for his marriage proposal: 'You can do it if you wish, Peter – but if you do, it'll be without me.'

Even prior to the ministry man's knock at the door many aspects of 'traditional farming' had begun to concern the couple.

They said no to the shiny, red fruit. If they'd said yes it would have meant many, many more birds in much less space and always indoors, other than on leaving the shed for slaughter. It would have meant birds crammed four to a tiny cage, in cages stacked up to six high. The birds would have lived, or rather survived, in faeces-filled, A4 paper-sized (or smaller), wire-metal units. The change would have meant painfully debeaking the birds prior to locking them in said cages. Debeaking was done because chickens in such unnatural conditions are unable to establish their natural pecking order and so often attack and cannibalise one another. This rarely occurs outside the battery and broiler systems.

Peter would have needed more medicines and antibiotics for his birds too. Intensively reared birds have a much greater propensity for diseases and

6 Kim Stallwood, *Growl* (Lantern Books, New York, 2014) pp.39–41.

disorders, including prolapsed uteruses, mouth ulcers, calcium deficiencies leading to brittle bones and frequent breaks, chronic respiratory defects, bodily lacerations and burning eyes from the high ammonia levels caused by the excess levels of chicken shit in poorly ventilated cramped barns. Peter and Anna reasoned if they intensified the chickens, in time they'd have to do the same with their cattle.

The hen battery system had begun in America in the thirties, and 15 years later Britain had followed suit. By 1947 the conditions that enabled factory farming in Britain were ripe. Post-war, with the population tired of never having enough, the government passed a new agriculture act and introduced farm subsidies. High levels of production were rewarded regardless of hidden costs and environmental toll.

It was American housewife and small-time chicken farmer Celia Steele who had created what would become the template for the entire modern battery and later broiler system.[7] Her story is one of chance and Frankenstein science, the kind of science that manipulates and degrades, rather than enhances, life. Mrs Steele was working on her farm deep in the Delmarva farmlands of Maryland in the thirties when she was mistakenly sent 500 chickens rather than the 50 she'd ordered. The rise of chicken hatcheries and artificial incubators had made the large order of chicks possible. She decided on a whim that rather than sending the birds back she'd keep them indoors and see if they lived through the winter. She put them in a cramped shed, feeding them on the newly developed supplements without which the birds would have died from lack of vitamin-rich sunlight.

In 1930, the average flock size in America was 'still only 23'; by 1935 Celia had '250,000' birds.[8] It was the peak of the Great Depression. Panic had followed the 1929 Wall Street Crash, vagrancy rocketed, the Missions expanded, and the breadlines trailed on and on. Alcoholics Anonymous was established as men who'd lost economic power and status tried to obliterate themselves. Ten years after her infamous 'breakthrough,' the Delmarva peninsula was the poultry capital of the world.[9] The welfare of a few thousand birds seemed far from the point – and Celia's discovery was a potential pot of gold.

Today in Maryland chickens outnumber humans a thousand to one, and its once beautiful and ecologically diverse Chesapeake Bay is a polluted mess thanks to the explosion of battery farms nearby and consequent pesticide pollution. The area is also a hotspot for outbreaks of avian flu. Current Compassion CEO and writer Philip Lymbery focused on the bay

7 Jonathan Safran Foer, *Eating Animals* (Hamish Hamilton, St Ives, 2009) pp.104–5.
8 Ibid.
9 Ibid.

in his explosive 2014 book *Farmageddon: the True Cost of Cheap Meat*. He interviewed resident and executive director of Chesapeake Waterkeepers, Betsy Nicholas, who stated that 'The three biggest problems with pollution in this area are agriculture, agriculture, agriculture.' She goes on to describe 'algal blooms ... and dead spots in the Bay'.[10] But as she says, 'local folks are reluctant to address the pollution issue because they don't want to be seen to be attacking family farmers. But factory farmers are a different matter. It's ... an industrial pollution source and really needs to be addressed as such'.[11]

Here is Peter with an aligned message, writing in the eighties in the Compassion in World Farming newsletter, by then called *Agscene*:

> Few realise that farm buildings enjoy freedom from rates ... all well and good in the old days of rotational farming when animals were related to the land, but today the position is quite different. Industrial livestock buildings house massive populations of animals which use imported feed stuffs and excrete slurry ... they are owned by companies quoted on the stock exchange with profits into seven or eight figures.

The devastation at Chesapeake is a microcosm repeated, to a greater or lesser extent, wherever the intensive farm reigns, and the residents of the bay are a microcosm of popular opinion. The pastoral arcadia is a hard myth to kill. Here is environmentalist and journalist George Monbiot writing on the 'Arcadian idyll' in 2015; he notes how:

> a conception of the shepherd's life, in both Old Testament theology and Greek pastoral poetry, [is seen as] the seat of innocence and purity, a refuge from the corruption of the city and it resonates with us still. But in the midst of a multi-faceted crisis – the catastrophic loss of wildlife, devastating but avoidable floods, climate breakdown – entertaining this fantasy looks to me like a great and costly indulgence.[12]

Yet even before the ministry man's visit the Robertses were experiencing Tom Regan's Muddlers phase in terms of their animals; the rural ideal had

10 Philip Lymbery, *Farmageddon* (Bloomsbury, London, 2014), pp.48–51.

11 Ibid.

12 *The Guardian* 22.12.15, https://www.theguardian.com/commentisfree/2015/dec/22/festive-christmas-meal-long-haul-flight-meats-damaging-planet

begun to lose its innocence for them. Anna was sleep-deprived, crying late into the night, expressing her worry to Peter over the fate of the unweaned calves he'd taken to market; she was of course herself by then a mother of two, so she naturally empathised more fully.

The cycle of birth, use and slaughter of the animals on a dairy farm followed a pattern that the Roberts were witnessing at first hand, and they were becoming more and more uncomfortable with it, especially as production pressure seemed to be intensifying. The heifers are usually impregnated at around fifteen months of age, either naturally with the bull or more commonly by being artificially inseminated, Gestation is for around 280 days (roughly nine months), then the calf is taken away from its mother at between one and three days old so that the cow's milk can be taken for human consumption. The male calves are either slaughtered for human or pet food or put into veal crates; more about those in Chapter 13. Cows are sociable animals which thrive in groups. 'It is thought that cows can identify 50–70 different cows.'[13] Then, typically 55 days after birth, the cow is impregnated again. This cycle continues until around the age of six when productivity declines below commercial effectiveness and the cow is slaughtered. The life-span of a cow in nature is twenty years.

Peter, considering himself a good farmer, did what he could to minimise the suffering implicit in this cycle by taking his livestock to the local Fareham abattoir for slaughter rather than make them suffer a long journey as a live exports to a continental slaughterhouses, where humane-killing standards were also less strict than in the UK. Unlike many farmers, he chose to stay with his animals to the very end, choosing to witness their suffering, inhale their blood and hear their cries rather as he said 'than feel the guilt of having just abandoned them'. In the end the barren cows upset him as much as the sweet-faced calves. After years of a cow giving 'her faithful best', he found it very hard, when her milk supply dropped below profitability, to smack her on the rump and say, 'Off you go, old girl.'

Twenty years on, writing a foreword to one of Anna's vegetarian cookbooks, *The Magic Bean*, Peter recalled this time in the early sixties recounting how,

> Anna and I, suddenly realised, with something of a shock, the depth of that unfaltering absolute trust which our farm animals put in us ... a trust which in the end we knew we could not honour unless we were to turn the farm into a sanctuary, for farm animals are bred to make a living for the farmer – in our

13 https://www.ciwf.org.uk/media/5235185/the-life-of-dairy-cows.pdf

case via their milk – and when they can no longer contribute they must die and the farmer must look away.'[14]

In this, the eighth year of farming at Little Barnet, Peter explained how they could look away no longer:

> It is hard that farming should be like this … Yet it has been so for thousands of years, ever since our ancestors employed animals to convert the fibrous grasses into meat and milk … that though in the main, animals have benefited from their association with man – that is while they are allowed to orientate to the natural world and while farming was a way of life … the age of technology has now relentlessly pressed the animal deeper and deeper into the production-line and exploitation [has taken] over from husbandry.'[15]

The Roberts had passed through Regan's Muddlers phase. As friend and ex-Compassion employee the late John Callaghan put it, 'they knew by this time that there were quite enough cruelties inherent in traditional farming without going into more intensive methods'. Peter described his Damascene moment simply: 'I considered the methods so cruel that all I could do was give up farming altogether.'

So the story changes here. The apple is not just refused but is turned into a magic bean, growing into a stalk which Anna and Peter climb, leading them higher, and to a new way of seeing the world.

Shortly after the visit from the ministry man the Roberts stopped eating animal flesh. Shortly after that they put Little Barnet farm up for sale and shortly after that began a rebellion from the back room of their new home. The witch was knocking on the wrong door!

Harking back to this so-called conventional start in life, the tagline for Peter on the Compassion charity website and elsewhere tends to read 'An ordinary man with an extraordinary vision'. Anna is missing from the slogan, not because she was any less a part of the vision but because even though she was campaign CEO her role was more behind the scenes; it had been her strong feeling against the cruelties of factory farm methods that solidified Peter's own

---

14 William Shurtleff, Akiko Aoyagi, *History of Soy in the U.K. and Ireland* (SoyInfo Center, 2015), p.1041

15 William Shurtleff, Akiko Aoyagi, *History of Soy in the U.K. and Ireland* (SoyInfo Center, 2015), p.1041.

doubts, and in the following decades it was Anna's hard work alongside Peter that made Compassion a leading light in the animal welfare world.

Beyond the potted tale, however, are the incremental epiphanies. In the seven years from the ministry man's visit to the official founding of Compassion in World Farming, Anna and Peter's consciousness grew. Just as all the human body's cells renew and change every seven years, philosopher and mystic Steiner also believed that every seven years important emotional and mental changes occur, in 'skin shedding' and 'transition in sevens'. According to Steiner's system, which Anna and Peter had an affinity for,

> from the thirty-fifth to the forty-second year, and in the next cycle from forty-two until forty-nine, a major change usually takes place. One begins to feel a new restlessness … a desire to share whatever one has gained through life with others comes to the surface … This is almost like unfolding something … this is when we reassess the results of what we are doing externally in our life. Our relationships, careers, habits and the ways we interact are all put under scrutiny and modified or changed. It's a time of facing up to what does and what doesn't satisfy us … People change partners, life directions, and even attempt major personal changes … In these years we move from old stereotypical roles with a new found confidence in our individuality[16]

Peter was 43 and Anna was 41 when they officially started the Trust that later became Compassion in World Farming, and over that seven-year period, from questioning the ethics of the ministry man's economics to the founding of the charity, the Roberts changed their diet, wrote countless concerned letters to various papers and existing animal groups on the issues surrounding intensive agriculture, attended environmental and animal welfare conferences and meetings, and joined a local spiritualist group with a strong focus on compassion. They sold their farm, abandoned their livelihood and re-thought their life's calling.

Though they'd chosen to say no to ministry man's suggestion and the wretched practices of intensification, others in the area were bowing to the economic incentives offered by factory farming. During this period Peter came across a letter in a Hampshire newspaper about a new broiler chicken farm in an AONB (area of outstanding natural beauty). The writer's objection was dismissed by the paper's editor, who claimed that only 'vegetarian housewives

16 http://dreamhawk.com/body-and-mind/every-seven-years-you-change/

from suburbia' were concerned with such issues. Yet this patronising and sexist dismissal of animal welfarists as sentimentalists with too much time on their hands – and who, furthermore, who didn't understand rural life – was *and still is* a common narrative. One which feminist writer and animal activist Carol J. Adams explores in her book *The Sexual Politics of Meat*, which focuses on the way that women and animals are both reduced to objectified body parts, and the way that men who show consideration for animals are often belittled, feminised. Utilitarian philosopher Peter Singer who wrote the landmark book *Animal Liberation* (1975) experienced a similar reaction when he first began advocating for animals. He noted how often he was asked, 'Why *animals?*' in a tone implying that animal welfare was surely only 'a matter for little old ladies in tennis shoes to worry about'.[17] Clare Druce, founder of the charity Chicken's Lib (1970), also recalls her mother and herself being referred to as 'just a couple of little old ladies' – Clare was 40 – in a *Sunday Times* article of the late seventies following their exposure of cruelty at a chicken battery;.

Peter wrote a letter back to that Hampshire newspaper stating, 'I'm a neighbour of yours and I don't quite fit into the category of vegetarian housewife from suburbia.' Here was someone with cultural authority, a working man, and an experienced farmer at that, who was daring to question the agricultural status quo. He followed up with the question, 'Isn't it about time you had a little bit of compassion for the flock?' His letter received huge support and the Roberts began to feel less alone in their cause.

In 1966, around a year before the official founding of Compassion, Peter attended an animal rights and environmental conference in London organised by Ruth Harrison, author of the 1964 ground-breaking polemic *Animal Machines*, which had received intense media coverage at the time of publication. In it she had called out intensive animal agriculture systems as absurd, noting how

> if one person is unkind to an animal it is considered to be cruelty, but where a lot of people are unkind to animals, especially in the name of commerce, the cruelty is condoned and, once large sums of money are at stake, will be defended to the last by otherwise intelligent people.[18]

Harrison was continuing along the path set by her father the famous Quaker, conscientious objector and writer Stephen Winsten, who had written books

17 Peter Singer, CIWF recording of speech at CIWF HQ, 5/6/14.

18 Ruth Harrison, *Animal Machines* (CABI, University of Oxford), p.175.

on two early animal welfarists, George Bernard Shaw and the humanitarian philosopher Henry Salt.

Harrison's work caused such a stir that it led to a UK government investigation into farm animal welfare, an investigation led by Professor Roger Brambell, the head of zoology at Bangor university in Wales and a Fellow of the Royal Society. He advised Parliament that animals should have simple and basic rights, including the freedom to 'stand up, lie down, turn around, groom themselves and stretch their limbs,' simple recommendations which became commonly known as the Five Freedoms. Throughout Compassion's two decades, Peter constantly referred to Brambell's report, frustrated that these commonsense recommendations had not been incorporated into law but rather remained as inadequate and ineffective 'codes of practice', ignored by many intensive farmers. Brambell's report did create some positive changes, though, and many high-profile animal groups adopted the Five Freedoms. In addition, the Farm Animal Welfare Advisory Committee was created off its back.

Harrison's conference was to reinforce Peter's own burgeoning views on the dire direction farming seemed to be taking. Peter Singer, reflecting on his own experiences at Oxford during the sixties, noted that although he was well versed in the other anti-establishment movements of the day, from women's liberation to the protests against the Vietnam war to anti-racism campaigns, it often seemed as if animals were nowhere among these ideas. By holding a conference of this calibre in 1966 Harrison was attempting to get animals more firmly established into the agenda of the Left. At the conference the key speakers, including economist Ernst Friedrich Schumacher, Cambridge professor and zoologist Professor William Thorpe and psychologist Dr David Cooper, introduced ecological, economical and humanitarian issues in an interconnected way.

Schumacher had been a protégé of renowned Bloomsbury set economist John Maynard Keynes, chief economic adviser to the UK's National Coal Board for two decades, and had become a figure in the popular consciousness when his book *Small is Beautiful: a study of Economics as if People Mattered*, hit the mainstream in 1977. He was an unusual economist, influenced by the practice of Buddhism, the writings of Gandhi and the philosophy of Thomism (the writings of 13th-century Catholic priest Thomas Aquinas). At the conference Schumacher argued that factory farming has no possible relevance to the world's hungry 'as the efficiency of husbandry dwindles by 80% in the production of meat'.[19] In *Small is Beautiful* he wrote

19 Internal CIWF document, CIWF Founding Ethos, by Carol McKenna.

> since there is now increasing evidence of environmental deterioration, particularly in living nature, the entire outlook and methodology of economics is being called into question. The study of economics is too narrow and too fragmentary to lead to valid insights, unless complemented and completed by a study of meta-economics.[20]

Meta-economics is the study of the very foundations of economics – money, the banks and the corporations – usually with a view to radical change of that system. Peter's bookshelves at his later home, Copse House, were crammed full of Schumacher's writings.

The writer and psychoanalyst David Cooper spoke next. Cooper was a friend and colleague of R.D. Laing, author of watershed book *The Divided Self*, and Peter was, like Laing, a member of the anti-psychiatry movement. To Laing and Cooper, going mad seemed to be the only sane response in an insane world. Before taking up private practice, Cooper had presided over Villa 21, a radical treatment unit for schizophrenics, set in the grounds of a larger, mainstream psychiatric hospital.[21] At the conference Cooper declared that 'when a man cuts himself off from nature around him he loses his yardstick of sound judgment and approaches madness.'[22] He argued that by putting animals into dark, comfortless cells and cages, humans are not only directly harming those individual animals but are also making psychic trouble for humankind itself. By the seventies however Cooper himself was off the rails, his methods as a psychologist both questionable and very of the time. There was little distinction made between himself as doctor and his patients, and so a strange intimacy of dysfunction was fostered. He prescribed LSD to patients. He ultimately came to a depressing end, dying in 1986 of chronic alcoholism. As a former, disillusioned, patient of his put it, 'the benefits of his wisdom seemed not to avail him personally.'[23] Yet speaking at the conference in 1966 – at that point still clearly lucid - he made some powerful statements, which resonated deeply with Peter.

Zoologist and ornithologist Professor Thorpe spoke next. He, a pioneer in the field of sound spectography, had created the first piece of technology that could analyse birdsong; he was also a leader in the field of behavioural biology. He explained that despite pro-factory farm claims that a distressed animal would not reproduce or feed, hence the farmed animals were not distressed,

20 http://www.centerforneweconomics.org/schumacher

21 https://www.theguardian.com/theguardian/2001/sep/08/weekend7.weekend

22 Internal CIWF document, CIWF Founding Ethos, by Carol McKenna.

23 https://www.theguardian.com/theguardian/2001/sep/08/weekend7.weekend

'that breeding and fattening are unreliable signs of contentment of animals in factory farms, as like forlorn human beings, unhappy animals may eat out of misery.'[24] He concluded that 'cruelty is any act that deprives the animal or bird of the power to respond to behaviour patterns built up over many thousands of years'. Factory farm systems are by default unnatural and often rely on deprivation in order to function at all.

Here, at this conference, Peter breathed 'a sigh of relief.'[25] He had finally found a kinship amongst these scientists, academics and animal experts, men saying what it was then very unfashionable to say. For Peter that kinship felt like the lifting of 'an enormous weight.'[26]

The times were changing and a year on from *Animal Machines*, writer and activist Brigid Brophy published her pioneering *The Rights of Animals*, serialised in the *Sunday Times*. Brophy, a lively writer, was a striking and sometimes controversial figure, in publicity shots seen smoking or holding a cat, and wearing zebra-print, metallic, or paisley. Considered an 'enfant terrible' of the sixties, she was openly bisexual at a time when this was uncommon, a pacifist, an outspoken critic of the Vietnam War, a feminist and an atheist. She wrote how

> the relationship of homo sapiens to the other animals is one of unremitting exploitation. We employ their work; we eat and wear them. To us it seems incredible that the Greek philosophers should have scanned so deeply into right and wrong and yet never noticed the immorality of slavery. Perhaps 3000 years from now it will seem equally incredible that we do not notice the immorality of our own oppression of animals.[27]

The title of her book alludes to three earlier texts: Thomas Paine's revolutionary and anti-monarchy *The Rights of Man* (1791–92), which gave rise to pioneering feminist Mary Wollstonecraft's *A Vindication of the Rights of Women* (1792), which in turn gave rise to classicist Thomas Taylor's satirical *The Rights of Brutes* (1792).

Taylor's satire had made the point that it was as ludicrous for women to have rights as for animals to have them: Brophy mooted that the passage of time would alter the majority's view on animal rights just as it had on women's rights.

---

24 Internal CIWF document, CIWF Founding Ethos by Carol McKenna.

25 Internal CIWF document, CIWF Founding Ethos by Carol McKenna.

26 Internal CIWF document, CIWF Founding Ethos by Carol McKenna.

27 Brigid Brophy, 'The Rights of Animals', *Sunday Times* article (1965) quoted in http://animeyume.com/animal_quotes/animal_quotes_industries.html

Anna, of Brophy's ilk, had asked Peter years prior and following the ministry man's house call: 'But don't the chickens have any rights?'

By the following year the Roberts knew they had to do something. They approached the existing animal charities of the day, asking them to take a position on factory farming. At that point there was almost absolute silence on the matter of farm animals. With the exception of a small group, the National Society for the Abolition of Factory Farming, the other charities, larger, more powerful, all responded to Anna and Peter's plea in the negative, some with derision.

Though Britain has a long history of animal welfare campaigning and can be described as a leader within the animal rights and welfare movement, the focus until the mid-sixties had been on pets and wild animals: creatures used for 'sport' or 'entertainment', not those for consumption. An incongruous emphasis, considering that the use of animals for food is above all else the most widespread cause of animal suffering in the world. As Peter Singer puts it,

> For most human beings, especially those in modern urban and suburban communities, the most direct form of contact with non-human animals is at a mealtime: we eat them.[28] ... The use and abuse of animals raised for food far exceeds, in sheer numbers of animals affected by any other kind of mistreatment.[29]

All options exhausted, and with the Damascene (or transformative) seed now firmly sown, Peter vented his frustrations with a sympathetic friend, Noel Earle Gabriel, a member of the spiritual group Anna and Peter had begun attending. Noel, a solicitor, responded to Peter's frustrations with, 'Well, Peter, you'll just have to do it yourself then. Come over to my offices tomorrow morning and we'll set up a trust.' Compassion had to be set up as a trust rather than as a charity because at the time – in 1967 – charities were not allowed to publicly criticise government legislation.

Anna and Peter did just that, and Compassion in World Farming (CIWF) was born.

28 Peter Singer, CIWF recording of speech at CIWF HQ, 5/6/14.

29 Peter Singer, CIWF recording of speech at CIWF HQ, 5/6/14.

4

COPSE HOUSE

Over these seven years of unrest in their lives as farmers in this newly intensified landscape and prior to setting up Compassion in 1967 a few major changes took place in the Roberts household.

In a perhaps somewhat unusual order, these changes occurred as follows: first, for a brief period they were oxymoronic vegetarian animal farmers, continuing to run the farm while they themselves were herbivores. They were of course dairy and chicken farmers, both of which involved animal slaughter – even under their free-range system. So after a while they quit farming animals altogether. Finally, they sold Little Barnet, and for the next three years, with toddler Judy and baby Gillian in tow, they rented various spots in Hampshire and Sussex. Life had become unstable and income insecure. Peter took work as a soil-tester for a local company whilst Anna looked after the girls, and the family tightened their collective belt.

Desperate for a bit more stability and somewhere to call home again, Peter spotted details for a house near Greatham in Hampshire.

'You won't be interested in that,' said the estate agent.

'No! That's *just* what we're after,' replied Peter. Copse House, was described as 'surrounded by well-known Selborne and other beautiful Hills and Scenery'. The village was well known because 18th-century curate, horticulturist and ornithologist Gilbert White had written his *Natural History and Antiquities of Selborne* (1789). He is considered one of the world's first ecologists and Peter was a big fan, so being in Gilbert White country was appealing.

But less so was the next line of the particulars, which read, 'The country is healthy and well adapted for Field Sports, and the Hambledon and Lord Leconfield's Fox Hounds meet in the neighbourhood'. After they had been in the house a few years, a terrified fox had run into the Copse House coal bunker and then into their garage where it hid behind the log pile. Peter had closed the garage door. The huntsmen had arrived and, hounds baying, loudly demanded that he open up the garage. 'You're on private property and if you don't leave I will be calling the police,' he had told them. They had no choice but to leave, and the fox escaped its barbaric fate … this time.

Copse House was a half a mile down a tree-bowered lane. At the top of the lane was the 19th-century church from which the lane took its name. Earlier it was called Pook Lane, and Anna often lamented the name change, preferring the more whimsical 'pook'. The house exterior was a page from *Country Life*: a whitewashed stone building, with climbing roses and wooden shutters at each window. Over the years Peter would paint the shutters alternately baby blue and bright yellow; he loved colour, even when mismatched and gaudy, a fact that his fantastic shirt collection testified to. (The best shirt in my opinion was a green, purple and brown one that featured design of pillars, nautical ropes and large medallions.) The large garden backed onto Great Woods, an ancient forest of towering trees and fallen mossy ones; in the spring the floor was lined with bluebells and wild garlic, and a large, still lake lay a few acres in. ('Don't go near the lake,' Anna would later warn us grandchildren every time we went for a walk, always fretting over our safety.)

Despite the glory of the garden and wood, the house inside was less *Country Life* centrefold and more building site. Two small cottages been knocked into one at the request of the previous owner, who had then decided to sell. It was in a chaotic state and condemned as unfit for human habitation, so the Roberts family got it at a good price. All the same, they loved the house and especially the woods and countryside surrounding it, and it remained their family home until their deaths.

After the purchase Peter set to work, usually with Frank, who came down from Birmingham where he lived as often as he could for days of DIY and evenings of whisky drinking – but in Frank's caravan; Anna didn't want them behaving raucously in the house.

The summer after moving into Copse House, their third child, Helen, was born. She was blonde and blue-eyed, with round, even features, the baby of the family, a family now of three girls, plus Anna and Peter and many companion animals. Around this time Peter, in an *Ecologist* magazine interview, identified himself as: 'both a farmer conversant with the ways of farming, and as a parent, concerned about the sort of world that [his] daughters will hand on to their children'.

Peter was a fun-loving but sometimes gruff father, and Anna an alternately tender and strict mother. The Roberts girls remember how if you were in Peter's way, leaning into his line of vision in the car for example, he would move you forcefully out of the way rather than just ask you to please sit back, and how if you began crying in a way he deemed affected he would respond 'Don't turn on the waterworks, toots.' He loved to blare hard rock at top volume, drink copious amounts of red wine, and sunbathe for hours. Anna was a harsher judge when it came to hedonism and rarely saw shades of grey, although she was overall most judgemental of herself; she worried if she watched too much of 'the box' (TV), felt immense guilt about having an extra glass of wine (Liebfraumilch topped up with lemonade), and called herself 'thoroughly idle' when she took a rare day off from her work at Compassion, Direct Foods or The Bran Tub.

Alongside the human family a host of animals settled into Copse House; even with the farm gone, creatures still remained integral to family life, and Peter's three cats from pre-Anna days – Battle, Murder and Sudden Death – were renamed at Anna's insistence. Battle became Blanche, and Blanche soon gave birth to a litter of white kittens, and each Roberts daughter was gifted one. Judy had Imp, Gillian Frisky, and Helen Snowy. In the garden they kept chickens and later dogs, and at one point a chinchilla lived in the shed.

Peter and Anna were both avid gardeners, a love inherited by Judy, and they kept a large vegetable patch and many fruit trees alongside the chicken coop. It was seen as a treat for the Roberts girls to be given the job of egg collection in the morning, the smell of hay and earth in the air. If the chickens had pecked and eaten their own eggs, as they sometimes did, Peter told them to mix pepper and mustard in the broken shells to try and put them off doing it again. Peter, inspired by Aesop's Fables – his references were frequently literary – named the rooster Chanticleer. When he talked of the progression from free-range to factory farms he would quote dramatic lines from John Milton's *Paradise Lost*, the animals crying, 'Farewell Happy Fields/Where joy forever dwells! / Hail horrors!' The largest of the chicken coops was to have a later incarnation as a small dog sanctuary, when Anna's friend Jean Le Fevre, a fellow animal-lover and activist, told the family about a dozen dogs who were going to be put down after being used by a local laboratory. The Roberts created a temporary kennel out of the coop while they rehomed them all – bar one, a little terrier called Kim, who they kept as a companion.

For the first couple of years Copse House was just a family home, a respite from the previous years of renting and regularly moving home with a young family; however, just two years after the purchase swathes of the house would be taken over and established as CIWF's first headquarters. It would remain so for the next decade.

Neither Anna nor Peter had any prior experience in running a trust or anything of that nature, but inspired by their vision they began to recruit a small but steadfast group of volunteers who would help them get their message out. Most of these were friends they had met at the spiritual group they had joined, the White Eagle Lodge. Today, fifty years on, the charity is an institution with a large international HQ, offices or representation in twelve countries, over 100 employees, and the ability to mobilise over one million voices for change around the globe. Back then it was quite literally a back-room operation, volunteer run and led; the early work was mundane and utilitarian, mainly revolving around envelope-stuffing. In those days, beyond niche circles, there was very little information about the issues surrounding factory farming in circulation – and of course there was no internet. Few members of the public knew what went on behind the doors of the then fairly new 'crate and battery' systems, and even less was known about the murky world of chemical fertilisers and insecticides that were dousing more and more of the British countryside, so the Copse House team spent hours upon hours printing and sending out educational and campaign materials.

During these early years of the trust their work revolved around education, by canvassing both the general public and concerned farmers, circulating petitions, and organising small-scale protests and marches against the practices of factory farming. Soon a large printing press was to dominate the family lounge, pamphlets consumed the kitchen table and boxes of petitions and vegetarian meal-mixes led up the zig-zag staircase.

Compassion's work dominated family life to such a degree that one day Gillian, finding that with all the philanthropic activity going on in her home she was being ignored, decided to take action. Eunice Watson, a friend and volunteer – and nurse – was there, stuffing envelopes, when Gillian complained of a temperature and was given a thermometer. When the adults' backs were turned she put it in some freshly brewed tea. Eunice took the reading: 'Oh my goodness, darling – you must be *very* poorly!' Had the reading been real, Gillian would have been dead.

Not long after this Eunice helped the family out by providing money for an extension to be built at the back of the house so that home and campaign life could have some semblance of distinction. Eunice's gift was one of space, and this new room – a deep red-carpeted oblong with large windows joined at the corners and with a double door opening on to the garden – was christened the Sunroom. The Roberts and Compassion were on a shoestring budget. Neither Peter nor Anna ever drew a wage from the charity – neither Peter as the official face of Compassion, nor Anna, as a full-time volunteer who was also on the board of directors from the founding until the late eighties. Years later, Anna and Peter bought Eunice a new moped to replace her ancient, battered one in thanks for her financial help.

By the time of my childhood during the late 1980s and 1990s the Sunroom had become a place for games and theatre. For us, the key feature and excitement of the room wasn't its history, which we didn't then know, but the bar and snooker table. It was a party room and playroom, where we set up elaborate Scalextric and Meccano sets, and created dance routines to songs by the Backstreet Boys and from musicals, 'Tell Me More' from *Grease* being a particular favourite.

Later, when the parties died down, the Sunroom's role became more of an ascetic one. A small gym was set up with a stationary bike and a rowing machine, plus a metal pyramid to meditate in. Coming over for an evening visit we'd find Anna, then in her mid-seventies, and her friends Polly and Betty cross-legged meditating, the geranium, sandalwood and vanilla of Nag Champa permeating the air, the soft plinky-plonk of New Age music emanating from the old stereo.

Prior to these incarnations however, the Sunroom was the official and original headquarters of Compassion for a decade. The Roberts had named the hub of all this activity for prosaic reasons – it had windows all the way round and was basically a greenhouse – yet I like to think also because it reflected the spirit of their cause; not dictatorial and absolutist, but enlightening and diffuse.

5

PROJECT 70

> There is no savagery in Nature equal to what man does in factory-farming, victim as he is to a false sense of values that puts economics before compassion.
>
> Peter Roberts, 1970

1968 was the year of revolt and its summer was the summer of love. Compassion itself had come to light in October 1967 just three months shy of that tumultuous and optimistic year. Anti-Vietnam War and civil rights protests were at their height, the women's liberation movement was in full swing and the seminal New York Stonewall riots, regarded by many as the first major protest on behalf of equal rights for homosexuals, were just one year away. It was the year Martin Luther King was assassinated and was the year of the Black Panther raised fist at the Mexico City Olympics. In Paris, students rioted against capitalism and rampant consumerism, and closer to home the Northern Ireland Civil Rights Movements held its first march against anti-Catholic discrimination.

Anti-colonial protests and resistance were ongoing, much of the decolonisation of Africa having taken place from 1950 to 1970. There was the first comprehensive coverage of the wide extent of famine in parts of Africa, crises that soon became a direct focus of Compassion's work. In January 1968, amid all this revolutionary fervour, Compassion launched its first major campaign, the optimistically entitled Project 70, which aimed to abolish, or at least begin the phasing out of, factory farming globally by the year 1970. Compassion's second ever newsletter, written late in 1968, stated proudly:

> We have growing support from people in all parts of the world for this project. Since our last newsletter, Spain, Trinidad and Barbados have linked themselves with it. We are promised most active supporters from South Africa … and France … Project 70 … must be worldwide.[30]

For these early newsletters Peter wrote the majority of the campaign prose, and Anna printed them on the large machine that had taken over the lounge. Project 70's approach was two-pronged: eradication and prevention. Eradication where factory farming had already taken hold, and prevention, staving off, in the so-called developing countries where factory farms did not yet have dominion. Project 70 focused heavily on the 85-page Brambell Report, condensed into the Five Freedoms, reasoning that if its codes of practice became law then most if not all factory farming methods would become illegal. The Five Freedoms, if followed properly – a reminder: freedom from hunger and thirst, freedom from discomfort, freedom from pain, injury, and disease, freedom to express normal behaviour, and freedom from fear and distress – simply did not allow any room for intensive farming. At the time of Project 70's launch these freedoms were being broken every day by common, standard and legal farming practices. At the time, calves in the UK and Europe were being intentionally fed unhealthy liquid diets in order to induce anaemia, so that the flesh would be more tender for the consumer, and were being kept in tiny, barren crates. Suckling sows were being kept in cramped crates which prevented them from turning round. Cattle and sheep were suffering long, cramped, dehydrating lorry and ferry journeys as live exports. Hens and chickens were being kept in overcrowded, disease-ridden cages. Slaughter methods regularly involved an assembly line method in which animals might be waiting for two days knowing they were next, undergoing forced fasting, electrocution, and for religious slaughter, non-stun slaughter.

Over the next two years Anna, Peter and their group of volunteers worked tirelessly sending out newsletters, taking direct action, filling petitions and writing letters to ministers and the press.

When 1970 came and went, factory-farming methods were still as far as ever from being phased out, even though that year was officially named the Year of Conservation, an admirable initiative which focused on the plight of wild and endangered animals as well as on the field of ecology. But it had no place for farmed animals.

30 CIWF newsletter, October 1968, p.3.

Project 70 did however make some modest achievements over these first two years of Compassion's work, by acting as a temporary dam to the progress of intensification, as a significantly galvanising campaign and as one that began to sow the seeds for change. As Peter eloquently wrote one year after the founding, 'nothing is ever wasted' and 'the seed of compassion is often sown in someone's mind with the sower being unaware of it and after a long while dormant it sprouts up into the light'.[31]

31 CIWF newsletter, October 1968, p.3.

6

RADICAL PACKED LUNCHES!

> A vegetarian is still regarded, in ordinary society, as little better than a madman.
>
> Henry Salt

Whilst Compassion gained public and political momentum, family life pottered on. Judy and Gillian attended Greatham Primary School, a 20-minute walk from home. Anna, Judy and Gillian trudged up Church Lane each morning for school with baby Helen in a Silver Cross pram and their beloved Kim perched on its hood, the girls carrying their vegetarian packed lunches alongside their notebooks and pencils. Attending school as vegetarians in the Home Counties wasn't easy, even in the apparently permissive sixties and seventies – and judgement and prejudice came as much from teachers as from peers. When Gillian and Judy were very small the family even went back to eating meat for a brief while because Anna feared the girls would be bullied; but it didn't feel right and they swiftly returned to a vegetarian diet. The girls did stand out, though, with their packed lunches, everyone else enjoying hot, meat-heavy dinners. The setting of the school was idyllic, nestled in the Hampshire countryside full of wildlife, yet the family were somewhat ostracized, their seemingly esoteric ideologies clashing with some members of the local small 'c' conservative community. There was a teacher who had taken a particular dislike to the girls, who openly calling them 'little snobs' on account of their 'odd' food choices, usually Marmite, Tartex pate and Nuttolene sandwiches; conformity was key, and packed lunches were considered radical. In the end the school forced them to have hot dinners, which, without meat, meant very, very little variety; they were given so much boiled cabbage that Gillian sometimes went home with her pockets full of the stuff, a shock for Anna when she went to wash the uniform.

While campaign life was intense during this early period, family happiness was always paramount for Anna and Peter, so when one day Gillian came home crying after another particularly bad day at school, Anna and Peter asked her if she'd like to switch schools. They settled on the local Catholic school, Our Convent of the Lady of Providence. Although they weren't Catholic they did have an increasing interest in spirituality and the school was perceived to be a 'good' one. The main appeal for Gillian, however, was the outfits; in the school brochure all the girls were lined up wearing the most gorgeous uniforms. In the summer it was thin, cotton striped blue dresses with straw boaters, and in winter thick, woollen navy dresses matched with felt hats. So she agreed; and Judy and Helen soon joined her at the convent.

Both Anna and Peter valued learning highly, Peter with his lengthy private school education and Anna with her great love of reading stunted at school. (She had loved English but had not moved up despite knowing she was bright enough, because the teacher, she believed, 'had it in for her'.) Vegetarianism wasn't common at the convent either. Animals were under 'Man's dominion' according to the catechisms, and in mainstream Catholicism animals were considered soulless, but the nuns and fathers had more interest in discipline than diets, so the girls were once more allowed their 'weird' packed lunches. Mass and confession were novelties for the girls, and for a while all went well; Gillian and Judy especially enjoyed taking the wafer and Gillian, like many young girls before her, fixated on the figure of Jesus. However, most of the teachers were an archaically strict bunch, and all three girls experienced the fear of a chalkboard rubber hurtling towards their skull. The rubber was thrown for all manner of indiscretions: a wrong answer, shoelaces undone, outdoor hat indoors ...

Despite trouble with authority, parts of their school life were fantastic. At playtime the Roberts girls enjoyed games such as kick the can. There was the annual summer fête, complete with coconut shy, tombola, and a white elephant stall, where bric-a-brac and junk were sold. They took school trips to see the Egyptian mummies in London and after a four-hour queue, which caused one of the girls to wet her pants, they eventually saw Tutankhamun's tomb. Peter had a lifelong obsession with Egypt and the Ancient World that middle daughter Gillian inherited. There was once a 26-mile country hike; during the walk the most terrifying nun heckled them, riding behind them on the back of Father Fagan's moped – renamed Father Fagend owing to his chain-smoking.

But weekends and holidays were best, filled sometimes with minor grunt work – letter-stuffing and stamping – for Compassion, but more often than not with walks up the lane and explorations into Great Woods. Gillian ran into the kitchen in a frenzy one afternoon and told Anna that among the flowers of Great Woods she had just seen 'a little lady in a purple dress sitting in the forget-me-nots.' Anna replied nonchalantly, 'Oh, yes, that's a fairy, darling.'

On Sundays there was the satisfying chore of Brasso-ing the fire-hood, bread oven and coffee pot. Afterwards time for a hot tin bath in front of the fire for each of the girls and then an episode of the family television drama series The Forsyte Saga. Something of The Good Life ambience remained, albeit with an animating political cause driving much of family life.

Every summer for years the Roberts family went to the White Eagle Lodge garden party. The Lodge had in no small part influenced them in their decision to change their diets, quit farming, and start campaigning. Now it has centres across the world, but its mother centre at Newlands is just 6 miles from Greatham. It is a magical place, sheltered from the outside world by thick hedgerows, where hundreds of bunnies hop about the grounds; at the centre is the Lodge Temple itself, a round white domed building with open wings of gold stretched across the entrance. Here the small Roberts girls would sit on the lawn, eating strawberries sprinkled with sugar, dressed up as woodland creatures. Beyond all this more frivolous side to the Lodge was its Aquarian age spirituality, which had affected the Roberts deeply. They were White Eagle Lodge members for 25 years.

For holidays, there were trips to North Wales in a VW camper, the three girls dressed in matching Fair Isle jumpers and cords; in their teenage years they developed more individual sartorial styles: Judy would channel Jimi Hendrix in velvet flares and leopard print; Gillian moved from tomboy to hippy-ish glam-rocker; Helen was the most classically feminine in floral prints and soft wool sweaters, white lacey scrunchies in her hair. One of their favourite things was going to see the 'tricklies', Peter's word for the small babbling streams so often found in the Welsh valleys. They built dams as instructed by Peter, sitting on the river-bed in his black Speedos, and stayed in primitive shacks with outhouse loos and only cold running water or sometimes in quaint, provincial B&Bs. In an interview with animal advocate and former Compassion employee Mark Gold, Peter had explained his original attraction to farming as a desire to 'roam wild on a thousand acres of Welsh hills.'[32] But now, as farming was no longer an ethical possibility for him, he altered this dream, to be realised in regular family trips to Wales.

When her older sisters graduated from school, Helen begged Anna and Peter to let her switch to Eggars Grammar, so she could be with her cousins,

32 Mark Gold, *Animal Century: A Celebration of Changing Attitudes to Animals* (Jon Carpenter, 1998), p.127.

Karen and Mark, who attended it. As it was next door to the convent Anna, who always took great pride in the close-knit nature of her family, agreed. Meanwhile Anna and Peter began to build a business with Karen and Mark's parents. Anna and Peter were just about getting by on Peter's pay as a soil chemist but they had a calling and an opportunity to start a business of their own, and knew that whatever they did it had to be aligned with Compassion. So in 1969 they set up a vegetarian food company, Direct Foods. Their business partners were Anna's sister, Jan, the one who had been the chaperone during Anna and Peter's five-day courtship, and Jan's husband Bob Howe.

Direct Foods was founded two years after Compassion. Like Compassion, it began around the Copse House kitchen table, although it soon spread throughout the entire house. The kitchen became the weighing room, the lounge packaging, and the rest of the house storage. Direct Foods sold soya-based vegetarian chunks, minces and meal mixes. The name Direct Foods was on the idea that it was wasteful feeding highly nutritious, human-edible foods to farmed animals when they could be fed directly to people. Direct Foods wanted to make becoming vegetarian and vegan easier, as well as providing options for those aiming to reduce their meat consumption.

Peter, having posed the question 'What if some procedure could offer a 500 per cent increase in food yields, with no danger of side-effects to ourselves or to the environment?' answered himself in that this technology already existed and that it meant 'going straight to the plant crop for our food instead of putting it through the digestive tract of an animal'.[33] Direct Foods was an ecological and humanitarian-minded business, and its values aligned closely with Compassion's campaign message, which was as much about the planet's health as about animal suffering.

Prominent early Compassion campaigns focused on world hunger and famine. A 1970s *Agscene* cover read: 'Save Grain for the Starving!' and the article inside exposed Western agri-businesses who were taking groundnut meal from famine-scourged areas in India to fatten Western cattle. Peter highlighted how 'the production of animal protein squanders nine tenths of the energy and protein value of crops'. In this issue *Agscene* also drew attention to a BBC documentary which, highlighting this iniquity, was narrated by Barbara Ward, economist and President of the International Institute for Environment and Development, later made a life peer. Into the eighties, Compassion focused its campaigns on Ethiopia, when history, sadly, repeated itself with grain being exported for livestock despite the country experiencing peak – and well-publicised – famine.

33 *The Bio-revolution: Cornucopia or Pandora's Box?*, eds. Peter Wheale, Ruth McNally (Pluto Press, 1990), p.201.

Although from the very start Compassion was not a 'vegetarian charity' but one that advocated for humanely produced meat, dairy and eggs, Anna and Peter via Compassion did always promote a diet that relied *less* on animal proteins and much more on plant. It's worth noting Anna and Peter's relation to veganism at this point: on and off they did consider veganism but they never followed through entirely. Anna only ate only minimal amounts of animal products; she drank soya milk but did eat some cheese and eggs, the eggs for a long time sourced from their own ex-battery hens that lived in the garden. Peter directly addressed the issue of veganism in a 1978 edition of the Vegan Society magazine. He describes his experience whilst still a dairy farmer of taking his barren cows to the abattoir for slaughter noting that 'What I saw at the abattoir made me turn vegetarian, which was perhaps illogical, since I should have given up milk.'[34]

Throughout four decades of Compassion fundraisers, balls and events, meat has never been served. In the mid-nineties, Peter was asked by a Compassion employee if high welfare meat could be served at the charity's bi-annual fundraising ball at a hotel in London's Park Lane. He responded, 'You may, but if you do I shan't be attending.' Despite this personal preference, Peter was principled yet never militant or didactic when it came to consumption. He maintained that Compassion's policy itself was 'anti-cruelty and not vegetarianism' and although 'to some people, but NOT TO ALL, the two may mean the same,' and that his own stance was 'a matter of personal choice and nothing whatever to do with CIWF.' He liked to say 'I don't tell people what to do; I give them the facts.'

In line with this when Anna and Peter had approached Linda McCartney, appealing to her for her high-profile support of the charity, Linda had declined, saying that she didn't want her name linked with an animal charity that was not pushing a wholly vegetarian agenda. She had additionally criticised Compassion for employing non-vegetarians in the office, but Peter maintained that this did not contradict Compassion's campaign message and when one of the early Compassion meetings was bombarded by other campaigners – including, as he put it, 'staunch vegetarian, anti-vaccination groups, and even a hot gospel or two' – he said, 'This has to stop,' and insisted that the meetings could not become 'a do-gooders' jamboree'!

Yet when specifically attacked over his and Anna's choice to be vegetarian and to raise their family that way, Peter had clear, strong responses. He referred to these critiques of vegetarianism as 'chestnuts to crack'.

The first was the belief that 'meat is necessary to health'[35] and that 'it makes Johnny big and strong!' Peter asserted that this was a 'strange idea when one

34 Peter Roberts in Vegan Society Magazine, 1978, p.6.

35 Peter Roberts, *The Ecologist*, December 1976, Vol.6, No.10, p.362.

considers how unsuited meat is to human nutrition, with its excess protein, its lack of vitamin C, its lack of calcium (unless you eat the bones as well), and its toxic uric acid'.[36] (Bone broth, by the way, was not in trend in the mid-seventies when Peter was writing.) He went on list some of the specific dangers that intensive farming itself had added to meat's nutritional inadequacy citing 'the modern vogue for rearing animals in a state of subclinical anaemia, giving the male animal estrogens ... [and] the prophylactic use of antibiotics which render the pathogens in the meat immune to medical drugs'. Peter concluded that this would strongly suggest that meat was in fact far from making one 'big and strong'.

The other chestnut is a perhaps less familiar to the lay person; it's the pernicious belief that 'farmyard manure is necessary for the fertility of the soil',[37] that in 'reality the soil just needs the wastes returned to it' and that, worst of all, 'it doesn't matter whether these are direct plant waste, animal waste or human waste'.[38] As noted, Peter's focus on the soil was always integral to his ecological ideals, a focus that is now, at long last, being recognised by the mainstream. In 2015 the 68th UN General Assembly declared that year the International Year of Soils[39] and in a nod to this legacy, current Compassion CEO Philip Lymbery wrote *Sixty Harvests Left: How to Reach a Nature-Friendly Future*, published in 2022.

So Anna, Peter, Jan and Bob set to work building this vegetarian business, Peter liaised with British Arkady who imported soya from America for its various omnivorous products. At that point soya was not available in the UK or in most of Europe, unlike now. He then got in touch with Brian Welsby from leading health foods company Haldane Foods, and employed him to mix the soya proteins with seasoning and vegetables to manufacture a vegetarian meal range. There were veggie-burgers, sausages, goulash and (family favourite) faux bacon Smoky Snaps. In the beginning, nearly all orders came via *Agscene*, which led detractors to claim that the Roberts had only set up Compassion (two years previously) in order to boost Direct Foods orders; an arduous and convoluted business plan had it been true.

In the foreword to *The Magic Bean*, featuring recipes using Direct Foods produce, Peter gave a potted history of soya, as it wasn't widely known then or indeed available outside large Asian supermarkets in London. The appeals of soya were, and are, numerous; pound for pound it holds more protein and less

36 Peter Roberts, *The Ecologist*, December 1976, Vol.6, No.10, p.362.

37 Peter Roberts, *The Ecologist*, December 1976, Vol.6, No.10, p.362.

38 Peter Roberts, *The Ecologist*, December 1976, Vol.6, No.10, p.362.

39 http://www.fao.org/soils-2015/about/en/

saturated fat than meat, as well as being economical and ecologically sound.[40] Genetically modified soya and deforestation issues were not prominent concerns in the years of Direct Foods, nor was the proliferation of studies claiming soya a dangerous food due to the bean's high phyto-estrogen levels. An alternative note on this idea from nutritionist Patrick Holford:

> recent controversy around soya can in part be attributed to powerful pharmaceutical companies who are loath to have any natural (and unpatentable) products reduce the need for drugs such as HRT for menopausal women, as soya is shown to balance oestrogen levels, and of course there are the dairy producers who certainly don't want trends to shift from milk to soya.[41]

Direct Foods was the first company in the UK to sell composite soya and vegetable foods. They showcased at exhibitions and trade shows across the country and the family helped out on the stalls: the Roberts girls, the extended family, and early Compassion employees such as Elaine Scheperel and Kim Stallwood.

Direct Foods began to build momentum, and Anna and Jan would use their weekends and evenings to package the bulk TVP (textured vegetable protein) into retail packs, packets of faux mince beef, ham-style soy, and natural (unflavoured) chunks. These were the type you soaked in stock and then simmered or fried, adding them to any meal an omnivore might eat such as bolognese, cottage pies, casseroles and quiches. Anna and Jan – exhausted, chattering, and slightly hysterical – would work late into the night, weighing until midnight and then lining the staircases with boxes to be posted out on Monday.

Jan and Bob's children Karen and Mark had sleepovers with the three Roberts girls while their mothers worked. Jan was fifteen years Anna's junior – a surprise baby for their mother Dorothy when she was in her forties – yet the pair were very close; Jan had gone vegetarian around the age of sixteen, influenced by both her sister and her own adolescent love of animals.

Direct Food packages were sent out across the whole of the UK as well as to Malta, Jamaica and Greece. The largest orders came from Jamaica, because of the vegan Rastafari movement in that country by observers of the Ital diet.

40 Peter Roberts in Foreword to Anna Roberts, *The Magic Bean* (Thorsons, Wellingborough), p.13

41 Patrick Holford, *The Truth About Soya*: https://www.patrickholford.com/advice/the-truth-about-soya accessed 21.10.16

One product made by Direct Foods that every vegetarian of the era remembers fondly is Sosmix. British Arkady already had a banger mix; a dry sausage mix made from soya but which incorporated pig fat. Peter requested that they create an alternative version in which they replaced the pig fat with vegetable oil; this became the much-loved Sosmix. It was marketed as 'The New Food for Campfire and Caravan' and 'the best part of a vegetarian fry up'. Sosmix was seen as 'a miracle food' and as scholar and animal advocate Kim Stallwood says it was 'a staple for vegan animal rights activists in the United Kingdom from the seventies onward … Many a hunt saboteur started their morning with a Sosmix sausage sandwich'. It also became the favoured food of the Woodcraft Folk on camping trips. (The Woodcraft Folk are the non-religious, more progressive, version of the Scouts.)

One recurrent problem during Direct Foods life was the Food Standards Agency. First, they complained about use of 'sos' in the Sosmix, saying this was unsuitable for a vegetarian food product, and claiming it was misleading as it sounded similar to the word 'sausage'. Peter responded that 'sos' did not refer to sausage as the FSA suspected, but to the old Cornish word 'sos' which meant tossing – one tossed the soya protein in one's hands in order to create the Sosmix. Direct Foods later took flak from the FSA again for their vegetarian goulash; the FSA claimed goulash, to be named as such, must have meat in it. But the most farfetched of the FSA complaints was when they objected to a product being described as 'vegetable' because it contained salt, a mineral. Of course almost every packaged product in the world contains salt as a preserver and flavour enhancer. It seems that the FSA had taken an institutional dislike to this upstart vegetarian company.

Despite all this, the produce gained in popularity and customers began asking for it in their local health food stores. Peter wrote in *Agscene* how some responses to Direct Foods had been 'wildly enthusiastic – some so wildly enthusiastic that we dare not print them for fear of being accused of making it all up!' Anna wrote three cookbooks featuring Direct Foods produce. The first was the biblical-sounding, *The Earth Shall Feed Us*, with its very Aquarian age subtitle, *Cooking for the New Age* (1976). Anna co-wrote this recipe book with Jean Le Fevre. Anna and Jean tested the recipes at Copse House and at Jean's home in Crowborough, in Sussex, and another friend, Joanna Hicks, a strict vegan and member of the Animal Defence League, completed the book with illustrations.

In 1984, Anna's second book, *The Protoveg Cookbook*, was published by Direct Foods and sold 8,000 copies – a *lot* for a niche cookbook in the eighties, or indeed at any time. The book was a collaborative, co-operative effort, Compassion supporters, friends and the public sending in recipe ideas which Anna tested on Peter and the three girls. Her third book, *The Magic*

Bean (1985) had a bright neon yellow and orange cover, with a photograph of tomato and protoveg-stuffed marrows and runner beans. Recipes included vegetarian classics such as mushroom stroganoff, pea soup and Christmas loaf as well as vegetable bobotie, tropical delights and shepherds surprise pie (the surprise being the lack of meat).

Now in the 2020s, vegetarians, vegans and flexitarians have a plethora of choice with brands such as Redwood, Fry, Quorn, Beyond Meat, Moving Mountains and Tofurky, and it's common to be offered non-dairy milks and vegan lunch options in mainstream cafes across the world.

There is also a strong Western reducetarian movement; this is 'the practice of eating less meat – red meat, poultry, and seafood – as well as less dairy and fewer eggs, regardless of the degree or motivation', an appealing concept for those not willing to follow an 'all-or-nothing' diet.[42] In 2017 the inaugural Reducetarian Summit was held in New York. Not only that, but other campaigns abound that aim to support the public in reducing their meat and dairy intake, including Paul, Mary and Stella McCartney-led Meat-Free Mondays, established in 2009, which focuses on the links between carbon emissions and meat consumption. There is also the hugely successful Veganuary campaign which began in 2014 and which works to make those sometimes shallow New Year's resolutions meaningful by supporting participants to go vegan for the month of January. Furthermore, journalist Graham Hill's 2010 TED talk 'Why I'm a Weekday Vegetarian', has at the time of writing had 2,967,530 views and in 2013 former *New York Times* columnist Mark Bitman published the book *Vegan Before 6pm* (VB6) which helps those looking to restore good health by adopting a flexitarian diet.

These days celebrity vegans are shouting loud. But back in the seventies, as Peter Singer recalls, 'there were no commercially made vegan milks, and Peter was a real pioneer in that field too'. Singer recalls visiting the Direct Foods factory in Petersfield in the early eighties and that Peter had then predicted, 'that with the way the population was growing, we would all be eating it before too long, because we wouldn't be able to feed the world otherwise'.[43]

Via the mouthpieces of Compassion and Direct Foods, and later the vegetarian health store the family opened, the Roberts were encouraging a reducetarian, or totally vegetarian or vegan, diet long before these ideas were fashionable and long before the term 'reducetarian' was known. The word was coined by Brian Kateman in his 2014 TedX talk. In a mid-seventies edition of *Environmental Health* magazine, a journalist reported that Mr. Roberts presents

42 https://reducetarian.org/faq/ accessed 24.5.17

43 Peter Singer, personal correspondence, 10/09/15.

> a closely reasoned argument for using proteins from plants for Monday to Friday 'convenience' meals and reserving animal protein for Sunday lunch and special occasions, and insisting that animals should be farmed in 'free-range' conditions to enable the animal to live a worthwhile existence before satisfying the needs of man. Mr. Roberts developed his case in sound ecological terms which were difficult to refute in the light of recurring food shortages and famine in the third world.[44]

Anna and Peter opined on more than one occasion that one day it would be those who ate animal flesh who would be looked on as the cranks, as opposed to vegetarians and vegans like themselves and their family – an idea which brings to mind comedian Simon Amstell's 2017 mockumentary, *Carnage*, set in the year 2067, when veganism is the norm, looking back at the strange ideas people had in the past about it being okay to eat animals and their produce. Peter himself questioned what the world would look like without animal agriculture and in reply to the question, 'Will animals disappear from the countryside?' he answered,

> Undoubtedly there will be some eccentrics who will still insist on killing an animal to eat its body, the cranks of the 21st century, but we will start to look upon our present breeds of livestock not as a source of food but as tools for ecological management. We shall learn to delight in animals for their own sake, rather than for our own.[45]

However, though Peter and Anna were certainly leading lights in terms of the promotion of a vegetarian diet it's worth pointing out that the Vegetarian Society was established more than a century earlier, in 1847, and the Vegan Society in 1944, and that both were founded in the UK.

In Peter's analysis of meat consumption and agriculture, he acknowledged humankind's omnivorous history but viewed humanity's current meat-eating habit as a temporary, rather than long-term or sustainable method of obtaining nutrition. In an *Ecologist* magazine debate in 1976, he wrote that 'we had resorted to flesh-eating in order to survive the ice-ages but that we should not however, assume that because of this that we can continue the

44 'Conscience in World Farming' in *Environmental Health*, February 1975, from scrapbook, page number missing.

45 Peter Roberts, *The Ecologist*, December 1976, Vol.6, No.10, p.362

habit of eating livestock'.[46] We should not assume this, he continued, because such an assumption would prove environmentally unsound. Peter lamented 'that to support all this factory farming, we are turning the once varied British countryside into a vast barley field'.[47] He saw the shift from 'the carnivorous customs of the cave' to the 'controlled grazing of livestock and more recently the landless rearing of farm animals' as potentially incredibly dangerous.

Many of Peter's arguments for the case for vegetarianism, particularly surrounding the idea of the ice-age crisis, seemed to have come from both his own deep thinking and perhaps from a 1956 book published by the UK-based Vegetarian Society, *Why Kill For Food?* by Geoffrey L. Rudd. He was secretary of the society and later the General Secretary of the International Vegetarian Union.[48] Rudd's book included chapters on the history of vegetarianism, on the ethical arguments for it, the health advantages of such a diet, the world economic perspective as regards this diet and on the topic of Christianity and vegetarianism.

In the world beyond Britain, Peter noted in 1976 that, 'on a wider scale, 370 million tons of the world's harvest is now fed to livestock, enough to meet the total combined needs of the population of China and India'.[49] To this mountain of grain being fed to animals Peter adds that soya from the USA and fish meal from Peru was added to the grain blend, an ecologically disastrous mix. Compassion CEO Philip Lymbery devoted chapters to this appalling practice in *Farmageddon* and *Dead Zone: Where the Wild Things Were*.[50]

Forty years on, the problems with intensive farming remain the same. As Peter saw it, 'omnivorism' should have been a 'temporary diet for an emergency' that was 'now over' and that humankind, 'in an ecological society, must be vegetarian and non-violent'.[51] In his characteristically jaunty way Peter chose to end his address to the *Ecologist* magazine experts not with didacticisms but with a quip: 'Let us be warned. There is only one primate apart from ourselves which persists in meat-eating and that is the baboon – and look what nature has done to him with his nasty characteristics, his truculent behaviour towards his spouse and to add the final humiliation, his purple bottom'.[52]

Compassion was still being operated from home, but Direct Foods, outgrowing the house, was moved to Liss, nearby. The demands of that business, with sales tripling around this time, also surpassed Anna and Jan's

46 Peter Roberts, *The Ecologist*, December 1976, Vol.6, No.10, p.36?
47 Peter Roberts, *The Ecologist*, December 1976, Vol.6, No.10, p.36.
48 https://ivu.org/members/council/geoffrey-rudd.html.
49 Peter Roberts, *The Ecologist*, December 1976, vol 6. No.10, p.361.
50 Peter Roberts, *The Ecologist*, December 1976, vol 6. No.10, p.361.
51 Peter Roberts, *The Ecologist*, December 1976, vol 6. No.10, p.361.
52 Peter Roberts, *The Ecologist*, December 1976, vol 6. No.10, p.361.

available time, and so at Anna's suggestion Jan's husband Bob took over as manager. Bob was a kind, quiet man, a vegetarian – and a football fanatic with a penchant for ping-pong and gardening; he won frequent awards for his prize vegetables. When the business continued to grow they began to search for an even larger place and found one in nearby Petersfield, close to the train station. It was an old motorbike garage which needed a lot of work and came with a caveat of council byelaws that insisted that the front of the building be used commercially. This caveat sparked the third aspect of the family's campaign. As Jan put it, 'The only kind of store they were interested in setting up was one that promoted their values' and in 1978 they opened their vegetarian health food store, The Bran Tub.

The wholesome and slightly frumpy cliché of seventies vegetarianism is clearly there in the name, but as well as connoting fibre, a bran tub is of course also a lucky dip at a local fête. In the tub various unknown and exciting objects would be buried in the bran, just waiting to be pulled out. In a conservative Home Counties market town though, The Bran Tub was sometimes seen as a little 'out there'. What hit you first on entering was the powerful cocktail of nutmeg, asafoetida and garam masala. At school other girls would tell me they hated going into The Bran Tub with their parents because it 'stank', but all the same I introduced my sleepover friends to houmous, Smoky Snaps, Nuttolene, gelatine-free cola bottles and Nag Champa.

Towards the rear of the shop was the weighing station where the spices, herbs, nuts and dried fruit were packaged. It was an old-fashioned health food store, where loyal customers shopped every week for years. Customers would describe the shop as an Aladdin's cave. Many were ecologically minded, or ethical vegetarians and vegans as well as health-aware omnivores and environmentalists. There were also shoppers who were none of these things but simply couldn't find some 'exotic' ingredient for a recipe anywhere else in the area. Generally they were the type who wanted to buy organic produce and wholefoods in bulk, and who wanted to refill their detergent bottles from the huge vats at the back of the shop, spilling gooey liquid all over the carpets. While refill and packaging-free food and domestic shops are now springing up all over the country's high streets, back then this was quite a novel idea.

The shop was a family-run business, firstly by Anna and Jan and later by Karen, Jan's daughter. Gillian started working at The Bran Tub part-time aged seventeen, and went on to run the store. For a while in the nineties I worked there too, as a Saturday girl: weighing, shelf-stacking, and serving on the till – and, between serving customers, reading books on nutrition, ayurveda, essential oils and juicing. (On Saturday nights I would have to scrub the yellow spice stains from my fingernails before going out, as the turmeric had stained them like a smoker's.)

In its early days The Bran Tub was an extension of Compassion's work. Anna and Peter had envisioned it as a sort of New Age Centre which would include books and crafts in addition to wholefoods. They believed that it would provide Compassion with personal contact with the public – much needed – and a lot of early customers were Compassion staff glad to have a place that was 100 per cent vegetarian so close to the office.

The manager of The Bran Tub, Kate Makey, started working in the shop in 1982 as a Saturday girl when she was just fifteen. She recalled how the Compassion campaign table crowded with leaflets stood by the till and how she could not avoid seeing it every day. One night after work she went home and told her surprised mother 'No More Meat!' She has been vegetarian ever since. The shop had a strong ethical ethos and was about more than just commerce. When cod liver oil capsules became popular in the nineties, sales reps repeatedly pushed them at Gillian, who was managing the store; they told her that The Bran Tub was losing literally thousands of pounds in business as there was so much publicity on cod liver health benefits at this time. But Gillian refused.

Eventually Direct Foods became a victim of its own success when British Arkady, realizing how popular soya 'meat' had become, threatened to put Direct Foods out of business by themselves selling similar products, widely distributed, in different packaging.

So in 1985 Anna, Peter, Jan and Bob made the difficult decision to sell Direct Foods. The Bran Tub still lives on, has expanded in size and range since 1978, and is now owned by Gillian. A Guardian obituary from 2006 stated that it is 'still one of the best independent health food shops' around.[53] Sadly, in October 2022 the Bran Tub closed its doors for the last time, a casualty of Covid and online food stores.

Peter and Anna choose to promote vegetarianism not only for the sake of farmed animals but because of their perception that Western meat consumption could be directly linked to international famine. In that mid-seventies *Ecologist* magazine debate, Peter stated that he and the other men on the panel could at last agree that there was 'no shortage of food in the world', but that 'there is starvation because of poverty and because of greed and because we devote the major part of the world's resources and its expertise ... to the feeding of animals instead of children'.[54] Peter predicted that 'if we continue along these lines, famine will increase on a scale never before seen' and that

53 https://www.theguardian.com/environment/2006/nov/23/obituaries.animalrights

54 Peter Roberts, *The Ecologist*, December 1976, vol 6. No.10, p.362.

this will be followed by 'the collapse of order and finally in war.'[55] This is the kind of dystopian future imagined in the novels of Canadian author Margaret Atwood, particularly the Oryx and Crake series, Octavia Butler's *New York Times* bestseller *Parable of the Sower*, and Mike Meginnis's *Drowning Practice*.

In campaign writings and speeches throughout the seventies and beyond the level of Peter's concern in terms of the preservation of life on our planet is clear; he insisted that 'We must get rid of the farm animal in the food-chain. It has become the cuckoo in the human nest.'[56] Cuckoos are 'brood parasites'; they lay their eggs in other birds' nests, and their young are brought up by the host birds, accompanied by the demise of the birds' own young. An eleven-day-old cuckoo will push any other eggs or chicks out of the nest, ensuring that it alone receives the attention and food from its adoptive parents. The factory farmed animal in Peter's analogy is thus described in wholly negative terms as a parasite, rather than a source of nutrition or a component of a healthy food-system.

55 Peter Roberts, *The Ecologist*, December 1976, vol 6. No.10, p.362.

56 Peter Roberts, *The Ecologist*, December 1976, Vol.6, No.10, p.361.

7

CEYLON TEA AND THE CAFÉ ROYAL

The Ceylon Lion and stamp of quality swung above the entrance of the Tea Centre, a stone's throw from the hustle of Piccadilly Circus. The Roberts' eldest daughter Judy and family friend Jeremy Hayward handed out flyers in the bustling streets. He recalls the first time he heard Peter speak at his school in Petersfield; the speech was emotive but fact-filled and a revelation to the majority of boys, who had never before given much thought to how the food on their plates was produced.

The leaflets teenage Judy and Jeremy handed out were simple A5 monochrome flyers that read 'Do They Deserve this Prison' and 'Is this Justice?' with photographs of pigs and chickens enclosed in barren crates, stacked roof-high. One flyer stood out clearly as having Peter's distinct voice, the gruff, unsentimental certainty; there was a photo of ten-day-old, un-weaned, dairy calves boarding a lorry, about to embark on the start of a journey as a live export which read: 'The Experts Say This Is Not Cruel. Any Damned Fool Knows It Is!'

Just down the road from Judy and Jeremy, Vogue cover girl and catwalk model Celia Hammond garnered support for Compassion by posing for photographers inside a scaled-up battery cage on Regent Street, emphasising the tiny space a chicken had under an intensive system. The press photographers had requested Celia 'sex up' her look from her original trousers and a T-shirt, so she had changed into Judy's velvet brown shirt, which she wore as a tiny dress cinched with a silver circle belt. Judy's feeling were mixed; excited that one of the most famous models of the era was wearing her clothes – but, at thirteen, slightly offended that her top had made a dress for a grown woman. The scantily clad woman in a cage did indeed garner attention from a lot of the Press, both broadsheets and tabloids – and having Hammond support

Compassion's cause was a huge boon. She had been modelling since 1960, had been a classmate of Jean Shrimpton (the world's first supermodel) and Joanna Lumley at the wonderfully named Lucie Clayton Charm Academy in Belgravia's Grosvenor Gardens; she had been photographed by the top fashion photographers of the era including Norman Parkinson (she was favourite model of his), David Bailey and Terence Donovan, and she was considered very much a part of the swinging sixties scene of fashionable London. That was until, after a decade of modelling, she had chosen to give up that career for good and instead to devote her life to animal welfare causes. It had been while modelling fur coats for magazines that Celia had had her own Damascene moment; she had always loved animals and she began questioning the ethics and supposed sophistication of fur. Her connection to animals grew stronger, and not long afterwards she rescued a mother cat and litter of abandoned kittens from a derelict building near her home in Hampstead. The seeds were firmly sown for what would become her life's calling. She remained a Compassion supporter for many years and went on to found the Celia Hammond Trust, which cares for stray, feral and unwanted cats and dogs in London and the South East.

Not long after that, Compassion held another high profile event in the capital at the glamorous Café Royal on Regent Street. Brilliantly gaudy, maximalist, dripping in gilt ornamentation, mirror-walled, with red-velvet banquettes and an excess of glimmering lights, the Café Royal showcased 19th-century Parisian opulence at its height. Its founders had fled their creditors in France in 1863 to set up the restaurant/bar in central London, which went on to boast 'the greatest wine cellar in the world' and to host everyone from Oscar Wilde and Virginia Woolf to Bernard Shaw and Aleister Crowley to Winston Churchill. In the seventies, when Compassion held their event there, it was a favourite haunt of David Bowie, Mick Jagger and Lou Reed, as well as being the spiritual home of boxing; Muhammed Ali, Frank Bruno, and Henry Cooper drank, dined, fought in the ring at the Café Royal.

That night under the chandeliers and classically painted ceilings of the Cafe Royal Compassion screened their hard-hitting documentary *Don't Look Now, Here Comes Your Dinner*. The name was a play on *Don't Look Now*, a 1973 Gothic thriller with Donald Sutherland and Julie Christie, based on a Daphne du Maurier short story. Compassion's documentary film featured thickly moustachioed and mullet-haired vet John Baxter. At the start of the film he flicks through a classic children's Ladybird book on the farmyard, but then, when he states how modern-day farming 'is a far cry from the pastures of such story books', the image changes to a scene of pigs shackled in tiny stalls, to (visibly) diseased battery hens, to calves in their narrow veal crates. Then the camera zooms in on a shot of a fry-up. 'Is this the price we're prepared to pay for our breakfast?' asks the voiceover. As Peter explained that night after the screening much of the footage had been obtained via the 'confidential invitation of the farm manager who expressed disillusionment and sorrow at the intensive unit he supervises'.

Next Baxter explains how humans are now building up immunity to life-saving antibiotics because of the indiscriminate use of these drugs on farms; *all* animals are routinely dosed, whether they are sick or not. Animals are dosed as a so-called preventive or prophylactic measure because the conditions are often so bad on intensive farms that they are either rife with disease, or are so squalid that if disease did break out it would spread much more rapidly than on a free-range farm. Extracts from the documentary were shown the next day on Southern Television, and the whole programme on three northern networks; it was a Yorkshire TV-backed film.

The film's primary audience was school-aged children. Its tone is melodramatic and aesthetics are dated, but the facts of the film still hold true. It's worth recalling that this film was screened in the mid-seventies and that today one of Compassion's largest current campaigns is still the antibiotic health crisis. In 2011 the then Director-General of the World Health Organization, Dr Margaret Chan, warned that we are unfortunately moving towards 'a post-antibiotic era, in which many common infections will no longer have a cure and will, once again, kill unabated'.[57]

This is precisely what is happening now, with the so-called superbugs. If the institutions had listened when Compassion first highlighted this impending crisis we would not be in the situation we are now in. Long-time Compassion employee Phil Brooke, who had worked with Peter in the eighties, recently recalled his memories of Peter, saying how he seemed to have the qualities of an Old Testament Prophet.

At the time the National Farmers Union (NFU) tried to keep 'Don't Look Now, Here Comes Your Dinner' out of schools, and in response Peter took part in a televised discussion to defend the film's content against charges of bias. His critics came from the Ministry of Agriculture, the NFU and a number of agri-businesses, including the Northern Pig Breeding Company and the British Poultry Federation. Peter pointed out that his team had filmed inside two chicken batteries and two calf-rearing units. In the case of the batteries, one had shown a high standard (according to government standards) of management and the other a low standard. Likewise, in the calf units, one was operating according to the Brambell Standard, and the farmer himself had the opportunity to 'defend the system'. Dressed in the white boiler jacket of

57 http://www.ciwf.org.uk/our-campaigns/antibiotics-health-crisis/ CIWF website, accessed 7.9.16.

the scientist, the factory farmer in the film claims that the battery system is preferable for the hens in terms of welfare because inside the caged system the hens are protected from the elements of nature and other dangers of the outside world. (Studies show that chickens survive a predator attack 90 per cent of the time in a wild environment.) The factory farmer's comments bring to mind psychiatrist David Cooper's statements from the 1966 London Roundhouse conference that Peter had found so enlightening: that by removing ourselves too far from the natural world we begin to lose our yardstick for sane and rational thought.

This focus on schools was always integral to Compassion's work, and Peter often posited that it would perhaps be on 'the educational front that the more significant progress would be seen' and he called factory farming, 'an explosive subject, and one which must be fully explored in schools'. He felt that an early dialogue with those who would later be running the commercial, social and religious life of the country in a few years' time was absolutely essential. Campaigning in schools did prove fruitful in a myriad ways. An *Agscene* report from the eighties covered a Somerset secondary school's exposé of their school's own farm; pupils from the school took an after-hours visit to their school farm unit, where they found the equivalent of 'slum conditions' and terrible cruelty. The photographs taken by the pupils showed 'a level of deprivation which is rarely seen in even the most intensive factory farms' and the publicity garnered from the story was enough to shut the 'Horror Farm' down.

The 2020s highly influential activism on issues around climate change and the environment, led by school-aged children and young people, bears testament to this view. But even back then, campaigning in schools did prove fruitful in a multitude of ways, one being that it exposed Philip Lymbery to the charity at a formative age. Like many others, he first encountered the charity during a school lecture. He was eighteen, and recalls that the man from Compassion explained to the students where their 'dinner came from'. Philip felt 'horrified by the pictures of pigs and calves in factory farms'. As a young man he kept homing pigeons and was generally fascinated by birds, so was particularly distressed by the images of 'hens in battery cages and how they couldn't flap their wings'. It seemed to him 'a crime' and on that day he 'resolved to do something about it'.[58]

The mid-seventies was for Compassion a time of spreading the word and using all media available, and so in 1975 they released a second documentary film, *Return to Tomorrow*. Peter had borrowed the name from an episode of *Star Trek* (he was a huge fan); his message was not subtle. In that particular episode

58 Philip Lymbery, *Farmageddon: the True Cost of Cheap Meat* (Bloomsbury, 2014) p.31.

the Starship Enterprise receives a distress call from a planet that had been written off as lifeless. In Compassion's 16mm film, the ecological and agricultural history of Britain is explored. The film begins by exploring how 'an ecologically stable system of agriculture evolved' whereby 'the farm animal was well-orientated on the land and the rotation of crops played an important part in controlling weeds, pests and plant-disease and in maintaining soil fertility'. The narrator goes on to explain 'how factory farming changed all that' [59] and how 'the once varied countryside grows barley year after year' and that a 'new habit of farming has developed in which toxic agri-chemicals play an important part. Pollution and the depletion of wildlife are the inevitable results.' The narrator probes, 'Have we got our priorities wrong?' and 'Do the laws of the stock exchange rule that we must devote the world's resources and our finest techniques to the feeding of imprisoned animals instead of to saving human life?'

Return to Tomorrow was released shortly after the 1972–73 Ethiopian famine in which 60,000 died, and the 1974 Bangladeshi famine, during which 27,000 people perished.[60] At the time of the promotion of the film, it was estimated that '10,000 men, women and children [were] dying from famine every day'.[61] The film concludes with the question, 'Must more die before we decide to cut back on wasteful methods of agriculture and make more available to the hungry?' and the statement that 'quality of life everywhere hangs in the balance'.[62]

With these pertinent campaigns on the interconnected issues of animal welfare, the environment and humanitarian crises, Compassion attracted further attention. Early patrons and high-profile supporters of the charity included conservationist of *Born Free* fame Joy Adamson; Russian Princess Helena Moutafian, a great philanthropist, who was also a patron of the NSPCC and Help the Aged, and the founder of the Anglo-Russian Children's Appeal; comedian and poet Spike Milligan; American author and founder of the International Vegetarian Union Woodland Kahler; and the great violinist Yehudi Menuhin. (At one point Gillian ended up with Menuhin's old chin rest for her violin. She learned a little but mainly played 'fatter than a caterpillar' scales.) Anna and Peter became friends with Spike and attended one of his weddings; through Spike Anna got to meet Harry Secombe, one of her heroes.

Of all the known Compassion supporters, Spike Milligan was especially adept at grabbing newspaper headlines. He was incredibly well-known in the country thanks to his ground-breaking and long-running comedy radio show

59 'Return to Tomorrow' article, 1975, source unknown. Personal scrapbook of Peter Roberts.

60 https://en.wikipedia.org/wiki/List_of_famines#CITEREF.C3.93_Gr.C3.A1da2009 accessed 20.7.17.

61 'Return to Tomorrow' article, 1975, source unknown. Personal scrapbook of Peter Roberts.

62 'Return to Tomorrow' article, 1975, source unknown. Personal scrapbook of Peter Roberts.

The Goon Show in which he, alongside comedians and actors Harry Secombe and Peter Sellers, aimed to ridicule 'the pomposity of those in authority and laugh... at the stupidity of mankind'.[63] Spike took petitions against battery cages to Downing Street, wrote open letters about the pitfalls of factory farming, and publicly protested. In the early eighties he protested the sale of pâté de foie gras in Harrods. It translates as 'fat liver' and is produced by force-feeding a tightly caged duck or goose corn through a tube two or three times a day, a process known as *gavage*. By the time that the bird is slaughtered for foie gras, their liver can be up to 10 times its normal volume, a painful inflammation which leads to hepatic steatosis. Along with the veal crate, foie gras is widely considered one of the cruellest of all factory-farming methods. It is now illegal to produce foie gras in the UK but it is currently legal to import and sell it.

The night before the protest, Carol McKenna and Joyce D'Silva cooked up £28 worth of spaghetti, the amount that a human would be forced to eat in a day were they under gavage. Spike, Carol and Joyce stood outside Harrods with a fake goose to show passers-by what happened under the gavage system, then the three of them carried the platters of spaghetti inside. There they politely asked to speak with the Food Hall Manager, but were denied. At the entrance to the meat section they were stopped by security. Spike offered to explain the visit but was ignored and they were asked to leave – but the protest had done its job and was aired on the lunchtime news as well as receiving extensive newspaper and radio coverage.

In the late seventies Spike wrote a letter to Peter for publication in *Agscene* (which he called 'Ag') summing up his views and his support for Compassion's work:

> Dear Peter,
>
> Let us ask a question. 'Why, up to the outbreak of World War Two was farming in the main, not on the battery operated system? Why should it suddenly burst out and get more intensive in the last ten years? The answer must be of course there is BIGGER DEMAND because (a) we are financially better off and (b) there are more of us.
>
> These cruel intensive methods can only operate if there is a market for them. Reduce the demand and you reduce the cruelty. Who makes the demand? The answer is WE DO. We, the people that is, you and me. It is no good somebody with a family of say

63 http://www.thegoonshow.net/

six children carrying around placards saying 'reduce intensive rearing' if they eat meat. The most economical and simple way to reduce intensive farming is for us to reduce our own numbers and intake.

So I would like to point out to everybody who is concerned with this, to consider the number of children they are going to have, in relation to the problem. You mention two meatless days a week. Well, if you can go two days why can't you go seven?

I myself am guilty of having four children, but I had not at the time woken up to the real problem. I am vegetarian and I try to influence my children, but they all eat meat and they eat chicken, which as we all know is produced under the most horrendous circumstances.

So in this respect I contribute towards factory farming four-fold. Had I known then what I know now, I probably would have had only one child. This not only applies to factory farming but it also applies to the whole quality of life on earth. The population of the world is increasing at an alarming rate, but the world isn't.

Even as I write, the last truly great rainforest in the world is being destroyed, to turn into a vast cattle-growing area. O.K. but one day there's not going to be any more Mato Grosso, then what? [Mato Grosso is a large state in west-central Brazil, at the time mainly covered with Amazon rainforest, wetlands and savanna.] In the last 300 years the world has filled up the last overspill areas on this planet, that is America, Australia, South Africa and New Zealand … staring us in the face is the obvious answer, that is to de-populate.

I am being very determined these days to speak the truth, and nothing but the truth regarding the quality of living in the world, and if there is to be compassion in world farming, people should give up eating meat permanently. I have done it.

I do hope that you and your readers might think seriously of this. Good luck to 'Ag', keep fighting. This is what I am doing, it's the only way.

Spike Milligan[64]

64 *Agscene*, Feb/Mar 1977, No. 44, p.11.

Even now I find Spike's claim 'Had I known then what I know now, I probably would have had only one child' rather startling, and while I doubt that he meant it to be taken at face value his typically shocking and direct mode of expression makes a strong environmental case for (non-fascistic) population management as well as encouraging those close to us to reduce meat consumption. Again, *Agscene* and Compassion's discussion points were somewhat ahead of their time, as it is only in fairly recent years that a more widespread discussion of the sensitive and complicated topic of population control in terms of climate change has taken hold in the mainstream media.

One day Peter got a call from Australian-born Peter Singer, then a 25-year-old student at Oxford. He is arguably now the most widely known moral philosopher in the world. His book *Animal Liberation* (1975) had an enormous impact, and 40 years on is still a handbook for anyone within the animal rights movements, be it campaigning against vivisection, factory farming or fur farming, or promoting a vegetarian or vegan diet. Now, in the 2020s, Singer is on the syllabus for every philosophy student.

It was a lunch break in 1969 that was to shape his life and lead him to Compassion and to Peter. One of his friends had ordered a vegetarian lunch, which in the England of that era was still, as Singer said, 'fairly unusual'. It led him, as it would 'a good philosophy student', to begin to research the animal question. He began reading and researching, moving from Aristotle's view that the 'less rational are there to serve the more rational' – a belief which, Singer noted, was 'used to justify slavery in Ancient Greece' – to Descartes' belief that animals are mere automata, 'complicated machines, like clocks.' Next came Kant, who believed only those beings with 'self-consciousness' were worthy of ethical consideration, every other living being merely 'a means to an end.' None of these satisfied Singer. Finally he came across what he described as 'a small footnote in [18th-century philosopher, social reformer and jurist] Jeremy Bentham's voluminous writings'. Bentham stated that animals at that time stand 'degraded into the class of things but that the day may come, when the rest of the animal creation may acquire those rights which never could have been withholden from them but by the hand of tyranny'. Bentham was writing at a time when slavery was still the predominant system in most parts of the world and only recently abolished in others; he continued his 'footnote', saying that

> the French have already discovered that the blackness of skin is no reason why a human being should be abandoned without redress to the caprice of a tormentor. It may come one day

> to be recognized, that the number of legs, the villosity of the skin, or the termination of the os sacrum, are reasons equally insufficient for abandoning a sensitive being to the same fate. The question is not, Can they reason? nor, Can they talk? but, Can they suffer? Why should the law refuse its protection to any sensitive being? ... The time will come when humanity will extend its mantle over everything which breathes.

In Bentham's writings Singer had found a philosophical framework he could align himself with. Bentham was equally appealing to Peter Roberts, and in campaigns he frequently returned to the idea of 'sentience as the only ideal which mattered' when it came to humanity's use of animals. (NB: the slavery/animal rights comparison is a problematic one, and one that is often used by white animal activists 'in order to further the narrative of animal rights'.[65] I recommend Christopher-Sebastian McJetters' essay on this issue, and the writings of A. Breeze Harper, aka Sistah Vegan, for a more in-depth understanding.)

Shortly afterwards a friend told Singer about Compassion in World Farming and he called Peter asking if he could have some leaflets for a stall that he and friends were putting on that weekend at the ancient Saxon tower monument in Oxford, close to the Ashmolean Museum. Peter did better than just bring leaflets; he arrived at the stall with a mock-up veal crate and a stuffed-felt veal calf as well as 'battery cages with papier-mâché hens inside'.[66] The calf, Victor, was doing the circuits of Compassion protests and events at the time. It had been a joint collaboration, made partly by students at nearby private school, Bedales, who were early advocates for Compassion's cause, and partly by Jan and Karen.

Singer's stall, with Victor, the cages and Peter's support, was a roaring success. Singer recalled that a lot of people were truly unaware of, and shocked by, the conditions the animals they were eating were kept in – and one passer-by, who apparently couldn't see very well, even complained of the students' 'cruelty at keeping hens in such tiny cages'.[67] Singer became a lifelong supporter of Compassion's work and in a condolence card following Peter's death in 2006 he acknowledged Peter's influence on his own developing consciousness.

---

65 http://www.sistahvegan.com/2015/12/28/the-prop-of-black-people-in-white-self-perceptions-revisiting-the-slavery-comparison-guest-post-christopher-sebastian-mcjetters/

66 Peter Singer, CIWF recording of speech at CIWF HQ, 5/6/14.

67 Peter Singer, personal correspondence, 10/09/15.

8

MYOPIC GENTLEMEN

Far removed from the glamour of the Café Royal, from protests and publicity in Harrods and from the support of Russian princesses and Vogue models was the daily grind of office life. Compassion's major focus in its first two decades was on the battery chicken, the veal crate, the dry-sow stall / gestation crate, and the cruel business of live exports.

On these four major issues Compassion educated the public. Revealing the realities of such intensive methods in the press, in institutions, on the streets and in exhibition halls, they also lobbied politicians and encouraged consumers to both reduce animal consumption and make more ethical food choices. Peter acknowledged that although 'farmers do not in general neglect or brutally treat their animals, there is another form of cruelty: deprivation.' [68]

Shifts gradually began to occur and in 1974 Compassion achieved a temporary ban on live exports in the face of strong pressure. However, as soon as the ban was brought in it began to be undermined by various groups and individuals, including the NFU and Lord Glenkinglas, the then chairman of the British Agriculture Export Council. Later that year Compassion encouraged MPs to support a total ban. An *Agscene* headline of that year reads 'Never mind the party – vote for humanity – in every sense of the word', followed by suggestions on how to lobby one's local MP. Peter encouraged supporters to write to their MPs regarding live exports but warned of those MPs 'looking for an easy way out who neither want to offend the farmers nor the town-dwellers' the 'town-dwellers' being largely anti-live export. These wavering MPs might latch on to the idea that live exports should continue but with 'maximum safeguards' in place – safeguards which, Peter noted, were (and

68 Source unknown. Unnamed article in personal scrapbook of Peter Roberts.

still are) rarely kept to and anyway are inadequate.[69] To make his point clear Peter suggested the reader should 'look up the safeguards which the Brambell Committee recommended and compare these with the pathetic and useless "codes of practice" into which they have been devalued'. Or, put another way, Peter said he felt that 'reliance on supposed safeguards did little more than "KEEP LIVE EXPORT GOING" by metaphorically plugging "THE LEAKS" to give the appearance of respectability, where truly there is none'.[70]

Beyond local MPs, Peter criticised the higher echelons of Whitehall including 'senior civil servants at the Ministry of Agriculture who cling desperately to the outdated concept of an export trade in live animals for slaughter' and who look upon 'parliament as their own public relations department'.[71] Peter encouraged the speaking of truth to power and of calling out careerist and corrupt politicians who seemed unable to engage with either logic or compassion. He questioned whether the O'Brien Report – a government report on the issue of live exports which led to debate in the House of Commons in 1974 – could be considered impartial, and whether or not its administration had been influenced by the Ministry of Agriculture.[72]

One might expect those working within the veterinary profession to be natural allies for an animal welfare charity, however time and again the opposite proved unfortunately to be the case, and Compassion called to account many vets employed to work on factory farms who propped up the squalid and cruel conditions for the sake of their own livelihood and profit. Following a veterinary inspection of an intensive unit in the seventies, vets had advised parliament that they'd found no evidence of pain or distress in factory-farmed animals. Writing in *Agscene*, Peter responded to this saying,

> The blind walk in light compared with those who WILL NOT see! The fact that these myopic gentlemen visited 4,000 units is entirely irrelevant ... What evidence will they accept? Is not the fact that the animal is kept in darkness sufficient evidence of cruelty? Or that it is unable to turn around? What has happened to us that we have suddenly abandoned common sense and call for objective evidence that we know will never measure subjective cruelty?[73]

69 Peter Roberts, *Agscene*, April 1974, No.26, p.1.

70 Peter Roberts, *Agscene*, April 1974, No.26, p.1.

71 Peter Roberts, *Agscene*, April 1974, No. 26, p.1.

72 Peter Roberts, *Agscene*, April 1974, No. 26, p.1.

73 Internal CIWF report, 'An Introduction to CIWF Founding Ethos', Carol McKenna.

Beyond letter-writing campaigns, petition-signing, and press coverage, Compassion also took a more active direct action approach, calling for 'activists who are willing to take part in non-violent demonstrations'.

In 1976 Peter and others took part in one such demonstration at the Royal Show, an annual agricultural fair held by the Royal Agricultural Society from 1839 until 2009. Peter described how he and other animal activists 'lay in wait' for a Mr Petrus Lardinois, the EEC farm commissioner. They were prevented from approaching him, first by show officials and then by the police. To solve this they used a megaphone to call him, saying that they were being prevented from delivering a letter drawing his attention to the fact that Denmark had kept a ban on battery-cage egg production since joining the Common Market, and calling for this ban to be extended to the rest of the Common Market countries. Peter called through the hailer for Mr Lardinois to visit the battery house in use at the show which was a 'disgrace to humanity.' Lardinois went over to the activists to listen to what they had to say, and admitted that he didn't like batteries either, stating, 'I hate the damn things.' But he also said that it although wasn't possible to extend Denmark's ban unilaterally to the EEC 'on economic grounds,' there 'is nothing in EEC law to prevent the UK from following Denmark's example.'

The activists spent the rest of the day stirring up some heated exchanges between farmers in the large gathering, and Peter noted that while two farmers were trying to justify battery farming, another turned on them, waving a Compassion leaflet and expostulating, 'How would *you* like to be shut up like that?'[74]

Peter was willing to stimulate debate between farmers on the ground as well as within the institutions which governed them, and in an article, 'The Flight From Nature', he wrote how the 'National Farmers Union, which saved the veal producers from legislation, and which was responsible for the re-introduction of Live Exports, has another black ace up its sleeve'[75] The black ace was of a seemingly more benign variety, yet Peter's criticism of the Shropshire NFU's decision to resume the cutting of road verges reveals the scope of Compassion's work beyond pure animal agriculture.

He went on to explain, in his typically pragmatic style, that although he is 'the first to agree with the clearing of sight-lines for the safety of road-users, previous reduction in unnecessary verge-cutting has resulted in the welcome return of many butterflies and other wildlife', and that it is 'good to remember that more than three quarters of all insect species are beneficial to man, and are in NO way pests'. Next, the influence of environmentalist Rachel Carson can be heard as Peter writes, 'how since the Second World War, intensive cultivation

74 Peter Roberts, *Agscene*, August 1976, No.42. no page.

75 Peter Roberts, 'The Flight from Nature', personal scrapbook.

and toxic herbicides have prevented any significant life in field-crops – even pastures', and that 'the insects' last refuge became the hedges and grass-verges, but then a fashion caught on which led councils to cut the verges every few weeks with rotary blades, destroying larvae, pupae & habitat.'

Peter's sentiments regarding hedgerow and declining vital insect populations were echoed in journalist and environmentalist George Monbiot's 2013 book *Feral*. He notes how farmers across the UK and Northern Ireland are having their subsidies threatened by allowing so-called 'encroaching vegetation' to grow on their land and that 'traditional hedgerows' are considered 'too wide.'[76] Such rules and subsidy cuts, as Monbiot noted, promoted, 'the frenzied clearance of habitats,' a system he suggested 'could scarcely have been better designed to ensure that farmers seek out the remaining corners of land where wildlife still resides and destroy them.'[77]

Back in 1976 Peter wrote presciently of the devastating effects of insect decline, and especially the approaching bee crisis; in that summer's *Agscene* he wrote of the Countryside Commission's concern over bee poisoning due to 'toxic chemicals sprayed on oilseed and rape crops'. The following spring he penned an article, 'Don't kill bees', describing how colonies of bees were being 'poisoned by pesticides.'[78] As is now widely known, bees are an absolutely crucial part of a healthy ecosystem; they pollinate everything from crops to fruits to wildflowers, with a third of the world's crops fertilised by insects and other pollinators.[79] Our food system relies on bee health. Today colony collapse disorder (CCD) is rife, and the bee population crisis – well, in the honeybee, *Apis mellifera*, at least; the fate of the thousands of other species of bees, many of which also pollinate plants, seems as yet to have gone unnoticed – is widely acknowledged. As was highlighted in London in 2017, at Compassion and the World Wildlife Fund's joint Extinction and Livestock Conference, due to this devastating decline crops in parts of China are now being pollinated by hand or by bee robots. As Compassion stated on their website in 2013: '**Global bee populations are in freefall, and one of the main reasons is the rise of neonicotinoid pesticide use on crops. Factory farming is partly to blame, driving a relentless demand for animal feed.**'[80]

At the time of writing (2022) numerous headlines have documented that the bee-killing pesticides known generally as neonicotinoids are making a comeback, thanks to policies by the UK government.

76 George Monbiot, *Feral* (Penguin Books, London, 2013) p.162.

77 George Monbiot, *Feral* (Penguin Books, London, 2013) p.162..

78 Peter Roberts, *Agscene* August 1976, No. 42 and *Agscene* May 1977, No. 45, p.9.

79 https://www.ciwf.org.uk/news/2013/06/the-end-of-the-birds-and-the-bees accessed 25.5.17

80 https://www.ciwf.org.uk/news/2013/06/the-end-of-the-birds-and-the-bees accessed 25.5.17

Peter's vision was one that placed its faith in traditional and more natural farming methods and he advised 'greater use of crop rotation and better cultivation in the control of weeds, crop-disease and pests could reduce the need for most agri-chemicals in use at present, and which do such damage to our wildlife'.[81]

Peter concluded his piece: 'most farmers are interested in preserving wild-life for its own sake but ... there is a self-destructive instinct in man which is associated with fear and a flight from nature.' This fear of nature's power and our true inability to control and dominate the natural world is 'expressed in a desire to kill any wild thing that moves, and cut down or spray any wild thing that does not move.' From the twee argument of the British need to preserve its iconic hedgerows, the debate moves towards a sublime and profound ecological theory.

Though Compassion's reach was growing during this era, funds, space and time were all stretched to their limit for at least the first decade of campaign life. The charity remained a backroom organisation, volunteer led and run, until the mid-seventies, and early *Agscenes* featured regular requests for grass-roots aid, both financial and in terms of people power. Then Peter issued a call for volunteers inclusive of all character types, introvert and extrovert, and through the gamut of professions, creative through to scientist. He wrote that although local active campaigns had steadily grown in strength and effectiveness since the last *Agscene*, and that this was heartening, they still had a long way to go, and Compassion needed much more support:

> Whatever your scene, whatever your skills and whatever your character, there is a way in which you can help ... Doctors, environmentalists or artists can all offer some application of their profession to the groups, whether you like standing up in a market place and shouting, or quietly handing out leaflets in a pedestrian precinct you still have a contribution to make, unique to you.[82]

For Peter and Anna, Compassion was a full-time job, juggled alongside the businesses of Direct Foods and The Bran Tub; in addition, in the early years Peter worked as a farm inspector as well as a soil tester, and Anna, as well as looking after the three girls, had a sideline in cooking vegetarian aeroplane meals.

81 'The Flight from Nature', Peter Roberts, personal scrapbook.

82 Peter Roberts, *Agscene*, issue 47, November 1977, p.11.

Seven years after its founding, Compassion required a general secretary, as Peter noted he would be devoting more of his spare time to writing, research and lectures. They were looking for a person who could provide 'regular attendance, on a part-time basis … capable of organizing meetings and events' and who, importantly, 'will be practical rather than emotional in his/her approach' because, as Peter bluntly phrased it, 'an emotional person would commit suicide in this job within a week'.[83]

Then, by 1976, a decade into campaigning, Compassion was able to afford offices, finally moving out of the Sunroom and, , like Direct Foods, into premises in nearby Petersfield. The new offices were in Lyndum House, a period building on Petersfield High Street, a prime spot. Petersfield is surrounded by farmed countryside and is a historically important area for sheep farming and cattle trading, and its square, which held regular farmers markets and is lorded over by an 18th-century statue of William of Orange on horseback, was a stone's throw from the new offices.

83 Peter Roberts, *Agscene*, issue 74, August/September. No page.

9

HAPPY WHIRLWIND

1976 was a dynamic and busy year. Compassion had rented a real office and hired paid staff for the first time, and had now been invited to be a part of the seminal Animal Rights Year. It began spreading its wings into Europe and aligning itself with various like-minded groups, including L'oeuvre d'assistance aux bêtes d'abattoirs (OABA, which in 1975 twinned with Compassion) and the World Federation for the Protection of Animals (WFPA), based in Zurich. Long-time friend of Compassion and animal campaigner Wim DeKok, who now works for World Animal Net, recalled meeting Peter for the first time in the late seventies at a small conference they had organised for farming welfare groups at Zeist in the Netherlands. Wim recalls how Ruth Harrison was at the conference as well as two elderly sisters (who Peter referred to jokingly as 'harum scarum', a description that stuck with Wim three decades on) from the German Verein gegen Tierquäälerische Massentierhaltung, now Provieh. That group, like Compassion and most of the welfare groups of the era, was small at the time, but was steadily and surely gaining clout, and it still campaigns on improving the welfare conditions for farm animals with a focus on 'making mandatory method of production labelling of meat and dairy products across Europe'.[84] Pre-internet, such connections were vital and Peter valued them highly. Wim, reiterating Peter Singer's sentiment regarding the lack of mainstream support for 'animal issues' at the time, stated: 'I have great respect for the pioneers who stood up for animals at a time when that was not popular'.[85]

Bearing witness to the huge scale of preparation for the Animal Rights year is a letter sent six years prior to the event from Peter to Tony Carding of the WFPA. Peter wrote:

84 http://www.labellingmatters.org/about

85 Peter Singer, personal correspondence, 10/09/15.

> Dear Tony,
>
> The Conference of' Animal Welfare Societies appointed a sub-committee to look into the possibility of an Animal Rights Year.
>
> Animal Welfare Societies have been asked about this and the general feeling is that if it is to be held at all it should not be too soon because of the enormous amount of work entailed in its preparation. The year provisionally suggested was 1975. However all have agreed on one thing and that it better to have no squib at all than a damp one!
>
> *If* sufficient interest is shown it might well be possible to approach the government (or certain foundations) for a grant, and to make presentations to the television, radio, press, magazines etc. for a greater than usual allocation of time/space during the year as their contribution to the same.
>
> Further it was suggested that an 'Animals Charter' should be agreed internationally … if we can focus public attention on the matter of Animal Rights in the way that it has been focused on Conservation in 1970 we shall take a great step forward. In the same way that the Conservation issue affected individuals personally so will the welfare issue be shown to affect them.

In the end, when the word 'rights' had begun to take on negative connotations, the Animal Rights Year became the Animal Welfare Year. Put together by the late Clive Holland, it included 67 animal welfare societies. Its two main focuses were vivisection (animal experimentation) and factory farming. As Peter mentioned in his letter, 1970 had been a vital year for conservation and he felt that this momentum could be capitalised on when it came to issues around farming. A 1976 ad placed in *The Ecologist* magazine read, all capitalised,

> IF YOU ARE CONCERNED ABOUT A SPECIES THREATENED WITH EXTINCTION, IS IT NOT LOGICAL THAT YOU SHOULD ALSO BE CONCERNED FOR A SPECIES THREATENED WITH MECHANISATION?[86]

86 *The Ecologist*, Vol. 6, No. 10, 1976, p.384.

The Animal Welfare Year, 1976, prepared the ground for the formation of a number of joint consultative bodies within the animal movement, including the establishment of the General Election Co-ordinating Campaign for Animal Protection (GECCAP) which, as Peter stated, presented 'a significant breakthrough in Animal Welfare, because it means that many different Societies are at last abandoning their traditional policy of "splendid isolation" in an attempt to bring about new legislation together'.[87] Owing to the GECCAP's campaigning, the UK's 1979 General Election was the first time ever that animals had a place in the manifestos of all the major political parties – a double boon during a time of intense political upheaval in Britain. The late seventies and early eighties were years of massive trade strikes; workers from nurses to gravediggers were taking industrial action, set against the backdrop of a bleak winter of blizzards and deep snow that led the press to brand it the Winter of Discontent, borrowing a line from Shakespeare's *Richard III*.

Compassion and the other members of the GECAAP campaign worked diligently, organising political conferences and exhibitions, and garnering attention from the Press. Compassion invited Liberal Democrat delegates to a meeting in which 50 signed a motion for an emergency debate on live exports. Compassion took a stand at the Labour Party exhibition at the Carlton Hotel in Blackpool, and received press in all strains of newspaper from the *Mirror* to *The Guardian*. In a mirroring of today's party lines, *Agscene* reported that the SNP (Scottish National Party) were sympathetic to their campaign and were 'particularly concerned about the competition to small farmers from large intensive units'. The Labour Party promised in their election manifesto to 'have stronger control on the export of live animals for slaughter, and on the conditions of factory farming', and they committed to a ban on 'hare coursing, stag and deer hunting.' The Conservative Party were of course the most reluctant to 'reveal their animal welfare policies'.[88]

During this intense season of campaigning, Peter made clear his own sympathies with the individual farmer, if not the institutions of farming and agriculture. He stated that 'it should be remembered that many intensive farmers would like to be more humane'. Speaking as a former farmer, Peter knew the economic and institutional pressures faced by many farmers. In recent years the high suicide rates amongst farmers have been in part attributed to both economic pressure and the predominantly conservative

87 *Agscene*, Oct/Nov 1978, No.51. p.2.

88 British Labour Party Election Manifesto 1979: http://www.politicsresources.net/area/uk/man/lab79.htm and *Agscene* 1979.

culture of farming that doesn't generally value emotional openness.[89] Peter went on to reveal that 'several [farmers] have told us how much they dislike batteries and pointed out that until the law bans them completely they simply cannot afford to do without them.' Peter was quoted in *The Times* explaining that Compassion had been founded by farmers, that they were not 'advocating the end of livestock farming' and that the 'platform is not vegetarianism. It is anti-cruelty'.[90]

Compassion employee Phil Brooke recalled a discussion with Peter in the late eighties regarding the vital importance of political engagement. Peter had explained to Phil, who had recently joined the organisation, that the charity Animal Aid were advising him that Compassion should be spending their time on the public, not lobbying politicians, and that Ron Brown, a Scottish Labour Party MP, wanted him to write the Labour Party's animal welfare manifesto – but it would probably take a lot of time and might not be worth it. He asked Phil what he should do, and Phil responded that Peter ought to take the time and write the manifesto. Phil said he had always felt that Peter really knew the answer to his own question, but wanted to get a feel for his new employee; Phil passed the test. Political action and public lobbying were not mutually exclusive, and like Animal Aid, he valued people power but also direct political involvement. He saw 'the modern equivalent of David's sling [in] ... the housewife's purse'! This was in the early seventies, and though the 'housewife's purse' comment might jar now, it was still common parlance in that era. (Despite the efforts of second-wave feminism, with publications like Betty Friedan's seminal book *The Feminine Mystique* in 1963, which had challenged the widely held belief that fulfilment for women could come only as housewife and mother, Peter's comment a decade on were still, disappointingly, socially accurate.) So Peter encouraged 'the housewife' to ask her butchers about the origin of the meat she was buying and the shops about the eggs they were displaying in hay-filled wicker baskets – what we'd now call eye candy, suggesting that the hens lived happy, outdoor lives when in fact they did not. Likewise, activist Lucy Newman of the National Society for the Abolition of Factory Farming (to become part of Compassion), is quoted during that era as saying: 'Our belief is that factory farmed products are much less hygienic and contain far greater hazards to health than naturally produced products ... [and] I know that housewives are aware of this problem.'[91]

89 https://www.pig-world.co.uk/news/more-then-one-farmer-a-week-in-the-uk-dies-by-suicide.html

90 Peter Roberts, *The Times*, London, 17.11.17.

91 Lucy Newman in 'Animal Lovers Keep Their Eyes on Farmers' by Hugh Clayton, *The Times*, London, 17.11.75.

For Compassion and like-minded groups it was a tale of David versus Goliath from the start; the agricultural conglomerates and NFU massively surpassed Compassion in funds as well as in terms of political clout. In early *Agscene* newsletters, the Roberts' fundraising pleas regularly come down to the basics, needing money to upgrade an almost defunct printing machine for their newsletters and leaflets. An early *Agscene* warned supporters that Compassion only has around 'nine months till we're out of business'. The charity had always been financially transparent, and even as late as 1983, when Compassion was much better established, Peter was able to make the pointed observation that 'The annual accounts of the National Farmers' Union are reported to show a £120,000 surplus over expenditure with the income of the N.F.U. from subscriptions amounting to £7 million per year in contrast to CIWF's annual income of £20,000!'

Yet the balance *is* shifting. Some corporations and multinationals normally considered natural 'enemies' of an animal welfare charity are making changes positively affecting the lives of billions of animals. There is no illusion that the corporate powers have undergone some kind of ethical enlightenment, yet the bottom line is that more animals suffer less as result of such changes. Over the last ten years Compassion's initiative, the Good Egg Awards, has seen numerous multinationals, supermarkets, universities and restaurants switch to higher welfare eggs and chicken. McDonalds has switched to 100 per cent free-range eggs in all their meals; Unilever-owned brand giant Hellmann have done likewise in their mayonnaises; across Europe, Lidl Germany now stock only free-range whole eggs ,as do Carrefour Belgium. Replace the word 'housewife' with 'consumer', and the sentiment holds true: 'When the housewife really wants to call the tune Goliath dances.'

As well as awareness-raising and political lobbying, the Animal Welfare Year led to a more intellectually unified approach within the various disparate animal groups, and in August 1977 the more aligned groups gathered at the ground-breaking Cambridge Rights of Animals symposium in the hallowed 16th-century Trinity College. The event, organised by the RSPCA, had speakers including philosophers Tom Regan and Stephen Clark, writers Brigid Brophy and Ruth Harrison, Labour politician Lord Houghton of Sowerby, psychologist Richard D. Ryder, Anglican Reverend and animal advocate Andrew Linzey, and animal campaigner Clive Holland. Peter went to the conference with Compassion's new campaign organiser and a recent employee, Kim Stallwood, who recalled the conference as a true 'eureka moment' when he felt as though he 'was hearing something which was articulating his own thoughts in a concise, clear and intellectually rigorous fashion'[92]

92 Kim Stallwood, *Growl* (Lantern Books, New York, 2014) pp.83–84.

The Cambridge Symposium upped the ante for everyone in the movement. Peter Singer regretted not being able to attend, but noted that 'it is not impossible when a century hence people ask where the newly victorious animal rights movement got started, historians will point to the meeting at Trinity College in 1977.' [93]

At the conclusion of the symposium, all 150 attendees signed a Declaration Against Speciesism. Richard Ryder had described and defined the prejudice known as speciesism in a pamphlet seven years earlier, and the term was later popularised by Peter Singer. Speciesism was added to the Oxford English Dictionary in 1985, where it is defined as: 'discrimination against ... animal species by human beings, based on an assumption of mankind's superiority'.

Yet speciesism is an ideology more complex than the one allowed by that dictionary definition, as it not only divides the human animal from the non-human animal, but also has numerous sub-categories within it: a hierarchy based on (perceived) intelligence and (perceived) levels of sentience, and a final term which disrupts and often contradicts the first two: the animals' usefulness to humans. In the dominance paradigm, pets or companion animals are at the top of the pyramid after humans; then endangered animals and the larger, more intelligent mammals are next; and the animals that appear to have a special bond with humans, like dolphins, gain extra credit. The arbitrary and Eurocentric nature of the hierarchy is apparent: South East Asian consumption of dogs and the strong Western condemnation of that is common – desperately hypocritical in light of all the evidence demonstrating that pigs are as intelligent and as sentient as dogs. And, sadly, the animals we eat are lowest in the hierarchy. The Cambridge Declaration Against Speciesism read:

> In as much as we believe that there is ample evidence that many other species are capable of feeling, we condemn totally the infliction of suffering upon our brother animals, and the curtailment of their enjoyment, unless it be necessary for their own individual benefit. We do not accept that a difference in species alone (any more than a difference in race) can justify wanton exploitation or oppression in the name of science or sport, or for food, commercial profit or other human gain. We believe in the evolutionary and moral kinship of all animals and we declare our belief that all sentient creatures have rights to life,

93 Review of *Animal Rights, a Symposium*: http://digitalcommons.calpoly.edu/cgi/viewcontent.cgi?article=1041&context=ethicsandanimals

liberty and the quest for happiness. We call for the protection of these rights.

For Peter to sign this was both a rational and radical act and by so doing he both aligned Compassion with other great thinkers within the movement, and solidified the academic and ethical framework from which the charity worked. The popular perception at the time was that animal charities were 'run on emotion and little else', but as noted in the 2006 *Telegraph* obituary of Peter, he 'brought intellectual discipline to his cause'.[94]

The signing was a radical act because most humans are naturally speciesist, or else are culturally conditioned to believe speciesism is natural, in that as 'human beings are more self-aware than other animals' and are able 'to think and act morally' this 'entitles them to a higher moral status'.[95]

Peter turned this argument on its head in 1984 in an *Ecologist* magazine debate in which he asserted that although 'There is violence in nature, as we are the superior species ... it is not our part to imitate, or to assume for ourselves the conduct of a pack of wolves ... motivation of the carnivore is not ours, [and in] man there is a quality which is unique, and it is the desire for justice.'[96] Critics of the anti-speciesist movement insist that this fight trivialises other more important social justice issues such as the fights against sexism, racism and homophobia. In response to this Peter asserted that 'the desire for justice' is the 'nucleus around which our intelligence was developed' and that evolution means the constant 'extending of the boundaries of justice'. He asked: '[Who would be] so narrow in his thinking as to say that our livestock and our wildlife are to be excluded?'[97] The battle against the speciesist ideology took the social justice movement to its logical conclusion.

That year at the Lyndum House offices in Petersfield, Compassion was able to take on its first paid employee, Elaine Scherperel, an American who'd moved to England the year before. Prior to this, Compassion had had just one part-time secretary, Pauline Parsons; she was the Petersfield-based 'practical person' who had met the criteria in Peter's general secretary advertisement. Elaine had met Anna and Peter via a mutual friend at a spiritualist group; she was being given board and lodging in exchange for babysitting duties but was looking for more substantial work in line with her ideology. Elaine felt that she appealed to Anna and Peter because she was a vegetarian, had a college degree, loved

94 http://www.telegraph.co.uk/news/obituaries/1535910/Peter-Roberts.html *The Telegraph* Peter Roberts obituary 4.12.06.

95 http://www.bbc.co.uk/ethics/animals/rights/speciesism.shtml accessed 7.4.16.

96 *The Ecologist*, 1976, p.362.

97 Ibid.

animals and could correspond in both French and German – a multilingual member of staff helped a lot with Compassion's European campaigns – and probably because she was willing to work for peanuts.

Elaine became Compassion's first 'National Campaign Organizer' a grand-sounding title for a job which in fact, according to her, involved a lot of sorting mail! Elaine would go through the ever-mounting pile and prioritise. She received rapid training, Peter filling her in on everything, and then began answering the plethora of letters from supporters, other charities and officials. Soon she was able to respond to 90 per cent of the mail without having to consult Peter, but the other 10 per cent was the challenge. As Peter only popped into the office at a fast run, she would follow him around while he attended to whatever he'd come in for that day. She recalled him as 'a happy whirlwind that blew in forcefully, brightened up the office for the time he was there, and then disappeared just as forcefully'.[98]

After a couple of months, Anna and Peter decided it was time for Elaine to start doing Compassion exhibitions and trade shows, so they bought her a small car along with some lessons to drive on the left. She began going to different towns to set up booths to talk to people about CIWF's purpose and goals and was also sent to 'meet people who seemed particularly interesting', which is how Elaine met Kim Stallwood, who was to become Compassion's second full-time employee.

98 Email correspondence with Elaine Scherperel, 27/09/2015.

10

COMPASSION

> Compassion is an unstable emotion. It needs to be translated into action, or it withers
>
> Susan Sontag, *Regarding the Pain of Others*

Kim Stallwood was twenty-one and newly vegan – by his own admission 'a vegan-gelical' – when he took his job interview with Peter in 1976, and he recalls that that moment was the first time in his life that he needed to think about the word 'compassion'. His mind at the time was, as he puts it in his own biographical account, *Growl,* 'a turbulent emotional mix of half-baked notions and outraged sentimentalities as regards animal rights and welfare'. He had dreamt of becoming a chef and had worked in several kitchens, but then he took a summer job at a chicken-processing factory. The decision to work at the slaughterhouse was a pragmatic one; he needed money, it was only for ten weeks, and many of his friends were doing it.

It was his experience there that complicated this seemingly straightforward choice. He had reasoned that as he cooked and ate chickens without thinking about them, then why not work where they were slaughtered? The abattoir in Hampshire employed 80 people, and each week it transformed 150,000 live chickens into pre-packed and frozen oven-ready birds. Kim worked on the post-slaughter section of the production line, where for eight hours each day he placed each carcass in a plastic bag. During those ten weeks, he could never bring himself to watch the birds as they were killed, nor to buy the oven-ready chickens offered for sale at a reduced rate as an employee benefit every Friday afternoon, even though he did continue to eat chicken. He struggled with his own inconsistencies, and a little while later, and following further debate with his vegetarian friend, stopped eating meat and soon after that went fully vegan. Shortly after this he met Elaine at the Compassion trade show stand.

Beyond the standard interview questions, about work experience, strengths and weaknesses, suitability for role etc was a question which Kim recalled over 30 years on; Peter asked if he had 'a problem with the word 'compassion', as some men are embarrassed by it.' [99] Kim swiftly said, 'Not me!' – but he admitted later that he had given that reassuring answer in order to get the job, and until that point he hadn't really given the word much thought. Kim got the job and worked for Anna and Peter at Compassion for the next two years as National Organizer.

It was Kim's first proper job, and during that time he helped established a nationwide network of local animal groups, organised campaigns and demonstrations, and helped with the production of *Agscene*. Among the more memorable trips of the era he recalls tailing a live exports lorry through France with Peter. The pair recorded onto a cassette tape what they witnessed, including a conversation with the veal calf farmer who felt that he kept his animals well, showing them that he had given them straw bedding even though he had put plastic bags over their mouths to prevent them from eating it in order to keep them as 'white veal'. Peter and Kim had ended the day by drinking gallons of *vin rouge*.

Peter's interview question to Kim was a revealing one, and the Roberts had chosen their campaign's name carefully; though Compassion in World Farming, was a far from catchy moniker, both Anna and Peter felt it encapsulated their ethos suitably and that this somewhat lofty focus on compassion must be central to their campaign designation. Peter's interest in exploring the deeper meaning of the word had sprung from his early interest in theology. Though worship was not an especially large part of his adolescence, his parents were both Christian in a loose sense, Peter's father Les seeming to have slightly more religious feeling than Peter's mother, Emmie. Peter's own interest in the mystical can be traced back to the 1940s, a teenage interest which took on an esoteric bent; at 15 he was a member of the Holborn-based Psychic Book Club, and he began practising yoga in his early twenties, roughly two decades before the Western yoga boom of the sixties and seventies. Importantly Peter's yoga practice was focused on the spiritual rather than only physical aspect of yoga and in Jon Wynne-Tyson's book *The Civilized Alternative*, Peter had heavily underlined the passage that read: 'the Asanas of yoga are as exercises meant only as a step to *Ahimsa*, which is compassion for all life, and *Karuna*, which is compassion in action.' Though he did enjoy the material aspect too, and was

99 Kim Stallwood, *Growl* (Lantern Books, New York, 2014), p.57.

often found doing headstands, or yogic inversions, against the walls of Copse House. He devoured yogic literature and particularly enjoyed Yogi Vithaldas' book *The Yoga System*. Vithaldas was one of the key yogis who had brought yoga to the West and was known for being a teacher to Yehudi Menuhin. It was yoga which led Peter to the sacred Hindu text of the Bhagavad Gita, the 700-verse Sanskrit poem set in the battlefield called Kurukshetra, an allegory of the ethical struggles of human life. In the Bhagavad Gita, '*daya*', translated as compassion, is contrasted with '*kripa*', pity. Unlike kripa, within daya there is a lack of otherness between sufferer and empathiser.

The Bhagavad Gita had a profound influence on Peter and it was a book he returned to again and again. He was to find an affinity with both the Hindu and Buddhist emphasis on compassion. In Buddhism the word translated as compassion is '*karun' and* similar to the Hindu term is understood as 'active sympathy or a willingness to bear the pain of others'.[100]

In his early twenties, Peter first read English poet and journalist Sir Edwin Arnold's eight-book blank verse poem *The Light of Asia or The Great Renunciation*, which tells the life story of Lord Siddhartha, who eventually became enlightened as Buddha. There is a strong focus in the poem on the necessity of mercy towards 'the beasts' and on renouncing 'the craving for flesh and blood', with the assertion that all animals 'crave life and have a will to live'.[101] Arnold himself was a vegetarian and the vice-president of the Bayswater-based vegetarian club that the Mahatma Gandhi had founded in the early 19th century.[102] In Buddhist thought, compassion and wisdom are generally viewed as joined roads towards enlightenment with the belief that 'Truly, you can't have one without the other'.[103] Thus inspired, Peter often spoke of the necessity of a balance between intellect and judgement, philosophising that intellect without judgement brought arrogance, that the converse brought sloppy sentimentality, and that either was wrong. Following this Eastern trajectory, Peter's study of theology led him to the ancient Indian Jain religion; as in Hinduism and Buddhism, compassion is a key component of Jainist belief and the life story of Lord Mahavira, the Jain leader, bears parallel with that of Lord Siddhartha (later the Buddha). Lord Mahavira, the last Jain who taught of God, became enlightened (*arihant*) only once he had conquered his inner passions such as attachment, anger, pride and greed, and had adopted

100 Barbara O'Brien, Buddhism and Compassion: http://buddhism.about.com/od/basicbuddhistteachings/a/compassion.htm accessed 7.11.16

101 https://ivu.org/history/europe19b/arnold.html

102 Hugh *Chisholm, ed.* 'Arnold, Sir Edwin'. *Encyclopædia Britannica.* **2** *(11th ed.). Cambridge University Press 1911*

103 Hugh *Chisholm, ed.* 'Arnold, Sir Edwin'. *Encyclopædia Britannica.* **2** *(11th ed.). Cambridge University Press 1911*

supreme compassion. The religion prescribes a path of non-injury (*ahimsa*) towards all living beings, and its practitioners believe that non-violence and self-control are the means to liberation, and that the function of souls is to help one another. In 1987 Peter received a Mahaveer Award at the Bharatiya Vidya Bhavan Indian Centre in West Kensington from the Jain group the Young Indian Vegetarians (YIV). Compassion received a second Mahaveer award in 1991. These awards 'are presented by YIV to individuals and organisations who have shown exceptional compassion for our animal friends'.[104]

Anna's family, like Peter's, were fairly nonchalant when it came to religion, and though Anna's mother Dorothy was remembered as being 'very kind-hearted' and, as Anna's sister recalls, was always looking out for 'waifs and strays and those in need'; this apparently came from a sense of natural charity rather than Christian obligation. Anna and her siblings Jan and Freddie had been sent to Christian schools 'because that was what everyone did at the time', rather than because Dorothy and Fred were great believers.

It was Peter who had introduced Anna to Eastern spirituality, and it was a little later that he was to discover, and introduce Anna to, the White Eagle Lodge. The Lodge melded various aspects of Eastern spiritualities with the teaching of the Christian Christ, as well as some aspects of Native American beliefs. Though esoteric, the Lodge has piqued the interest of some high-profile followers over the years, including the creator of Sherlock Holmes, Sir Arthur Conan Doyle. He held a firm belief in the existence of the after-life, researched paranormal activity, wrote a mostly ignored *History of Spiritualism* in 1926 and towards the end of his life was in correspondence with the White Eagle, founder, Grace. It was Lodge member Noel Earle Gabriel who had pushed Anna and Peter to start Compassion. Eunice Watson, who purchased the Sunroom for Anna and Peter, was a Lodge member, as were many other early supporters and volunteers. The first Compassion trustees were Lodge founders Grace and Ivan Cooke. In the eighties Lodge minister, Jeremy Hayward, Grace and Ivan's grandson, became a trustee of Compassion, which at the time of writing he still is.

The White Eagle Lodge was founded in Burstow, Surrey, in the early thirties, and more formally in Pembroke Hall, Kensington, London, in 1936. The teachings of the White Eagle Lodge are described as 'simple and pure ... built on a loving family spirit of helpfulness, kindness and sympathetic understanding of each other's difficulties, not only the practical difficulties, but the soul struggles.' The aim of the Lodge's teachings is to achieve 'a way of life which is gentle and in harmony with the laws of life, involving the belief that God, the eternal spirit, is both Father and Mother, and that the Son—the

104 http://www.youngindianvegetarians.co.uk/mahaveerawards.html

Peter as a young man.

Anna ready to ride.

Anna in her army uniform.

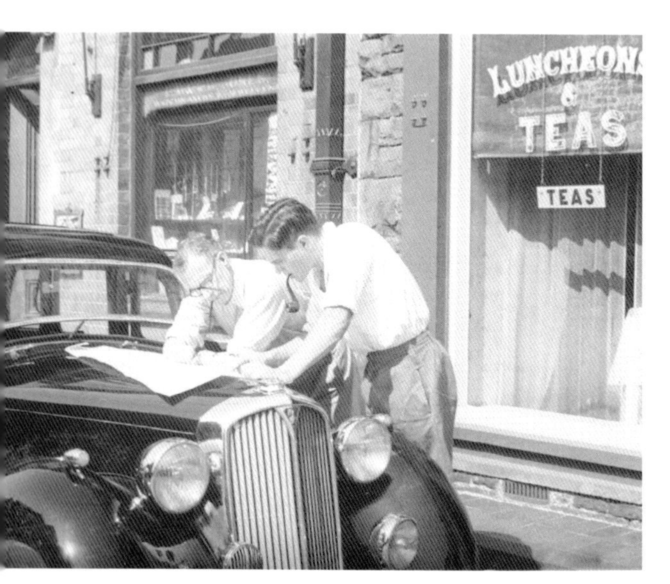

Peter with his father.

Mule in lady's hat on ship from Asia to Greece.

Peter outside the front of Harper Adams Agricultural College.

Anna and Peter stepping into their wedding car outside the church.

Anna and Peter raising their first toast as a married couple.

Peter as a young man climbing an oak tree.

Copse House, the headquarters of CIWF for the first ten years.

Anna, Judy, Gillian and Helen on a family holiday in North Wales.

Former vogue model Celia Hammond modelling in an upscaled battery cage for CIWF in 1970.

Eldest Roberts daughter Judy, left, as a teenager, caged in central London protesting against battery cages for CIWF.

Peter, right, speaking to a farm minister about a hoped-for ban on battery cages, mid-1970s.

CIWF protest outside parliament in Rome with Hetty the Hen in tow.

Hetty the (battery) Hen as created by the Spitting Image team for CIWF.

CIWF protests against factory farming down London's Park Lane.

CIWF live exports march in Brighton.

Peter talking about animal welfare at a church in Dover.

Spike Milligan delivering petitions and paper-mâché hens to No. 10 Downing Street on behalf of CIWF.

Spike Milligan with Peter at his side campaigning against factory farming of hens.

British actor and animal activist Virginia McKenna delivering petitions against factory farm methods to No. 10 Downing Street.

A CIWF protest outside the National Farmers' Union HQ.

Peter and Anna in the garden at Copse House.

Anna and Peter and former CEO Joyce D'Silva at the Animal Awards in the 1990s.

The author as a child.

Anna and Peter and friends at Sai Baba's ashram in northern India.

Peter on his tractor in his rewilded field.

Anna and Peter enjoying a game of Jenga at Christmas time!

Anna and eldest daughter Judy putting some
coins into the CIWF tin on Christmas day.

Cosmic Christ—is also the light which shines in every human heart'.[105] The Lodge's monthly magazine is called *Stella Polaris*, the North Star, one of the brightest seen from earth. The Lodge's symbol is a six-pointed star, known as the 'Christ Star', which is believed to 'radiate peace and healing to those in need.' The Lodge 'draws on the Ancient Wisdom contained in all the major religions of the world, and its keynotes are love, tolerance and service to all life.' [106] It combines elements of the religious teachings that already appealed to Anna and Peter, including Hinduism, the New Testament teachings of Christ (ideas like 'Love in Action') and aspects of Native American beliefs, linked to Gaia (embracing respect for the Earth as mother). Though in many aspects admirable, the teachings often lose more cynical individuals in light of their method of spiritual delivery: Lodge devotees believe that a deceased Native American named White Eagle imparted his wisdom via psychic messages which he delivered through or to Lodge mother Minesta (also known as Grace Cooke). White Eagle is described as 'a spokesperson for a group of illumined souls in the inner world known as the Star Brotherhood' and key amongst his messages is the spreading of 'light', a belief in healing, a belief in the afterlife and the importance of following a spiritual path, known as 'unfolding'. There is an emphasis on meditation and yoga, a belief in astrology, and on pagan and First People's aligned respect for the natural world.[107] White Eagle was said to promote, via Minesta, a 'great respect … for animal life', and he asked his followers to 'work for an end to cruelty, and for harmony in and with our natural world'.[108] Key to the Lodge's teachings is 'the promotion of respect and compassion towards all animals and sentient life'.[109] In a 1969 edition of *Stella Polaris*, Lodge members are encouraged to support Compassion's work. The article reads:

> Compassion in World Farming … is working to abolish the needless misery of factory farmed animals and, in the long term, to carry out an educational program for the development in all people of an attitude of kindness and respect for all forms of life.

The Roberts were official Lodge members for 25 years, although in later years Anna was a much more active member than Peter. He was just as

105 http://www.encyclopedia.com/philosophy-and-religion/other-religious-beliefs-and-general-terms/miscellaneous-religion/white

106 http://www.whiteaglepublishing.org/

107 http://www.whiteaglepublishing.org/

108 *Stella Polaris*, 1970.

109 https://www.ciwf.org.uk/about-us/our-trustees/jeremy-hayward-vice-chair/

interested in what atheist philosophers and irreligious writers had written on the concept of Compassion as he was in any scripture. He would quote the 18th-century radical French philosopher Voltaire from his essay 'Of Evil and in the First Place the Destruction of Beasts': 'We soon cease to be touched by the awful destiny of the beasts that are intended for our table … Children who weep at the death of the first chicken they see killed laugh at the death of the second' or early 20th-century philosopher, pacifist and vegetarian Bertrand Russell: 'There are certain things our age needs … It needs compassion and a wish that mankind should be happy … If you feel this you have a motive for existence, a guide in action, a reason for courage and an imperative necessity for intellectual honesty'.[110] The word 'compassion' was at the centre of the Roberts' cause and they both felt keenly that the next stage of human ethical development would be, as Peter put it 'to show that such things as compassion and justice are not restricted to our own species alone'.

110 'When Money Supersedes Compassion', *The Camden, Holborn & Finsbury Guardian*, 20.09.74.

11

WAVELENGTH

'They're ... you know ... on our *wavelength,*' was Anna's shorthand for people with similar philosophical, ethical and/or spiritual inclinations as herself, and she used the phrase broadly to describe people who would do something as simple as wearing organic cotton, drinking herbal tea and collecting crystals to those running serious humanitarian campaigns like theirs. In an early *Agscene* Peter described the more serious aspect of this 'wavelength', saying 'we are making links with other sections of the animal welfare world in order to exploit the advantages of one big, united army striving for one cause'[111]

By the mid-seventies Compassion's 'army' had grown significantly; patrons included actor Peter Cushing, most famous for his sinister roles in the gothic Hammer films of the seventies and his portrayal of Grand Moff Wilhuff Tarkin, commander of the Death Star in Star Wars; Malcolm Muggeridge, British journalist, satirist and former WWII government spy, known for helping bring the Catholic saint, Mother Teresa to popular attention in the West (as well as being seriously critical of Monty Python's *Life of Brian*); and Keith Michell, the popular Australian actor known for his multiple TV and film portrayals of Henry VIII. Peter was especially pleased to gain Cushing as a patron, and as someone who sometimes visited the Compassion offices in Petersfield because he'd admired the actor's work since Peter was in his twenties and seen him perform in a number of Shakespeare plays. Muggeridge had been a connection made at one of the spiritual groups the Roberts attended in this era, and Michell ended up becoming a friend of theirs who they'd go to watch perform in plays at Chichester Festival Theatre. As we've seen from the 'housewife' comments earlier, in this era the image of animal campaigners was often a clichéd one,

111 Peter Roberts, *Agscene,* 1968, Winter Issue, p.3.

full of prejudice, so such high-profile names gave the charity legitimacy and respectability as well as greater exposure. The mid-seventies image of the 'typical' animal campaigner is summed up by an article in the *Environmental Health* magazine which was profiling Peter, and which noted with surprise that 'Mr. Roberts is not the emotive and rather cranky speaker one comes to expect from animal welfare pressure groups although he believes deeply in his work and has developed his philosophy to an advanced level.' [112] As Peter Singer recalls, it was 'not a joke that vegetarians were thought of as cranks at that time, as the leading vegetarian restaurant at the time, in a pose of self-mockery, was called Cranks.' It was a Covent Garden-based wholefood vegetarian and vegan restaurant which enjoyed a successful run of over 40 years as well as publishing a host of Cranks cookbooks, which were amongst Anna and the whole family's favourite recipe books.

Beside actors and writers, Compassion had from its earliest days attracted members of the nobility on their 'wavelength' who became part of their 'army'. Their early patrons included aristocratic patrons who became very involved in Compassion's work , such as Lord and Lady Dowding. Like many early supporters the Roberts had met Lady Muriel Dowding and her husband Hugh via the White Eagle Lodge. Lady Dowding was led by a strong spiritual philosophy that she'd inherited from her mother, Hilda Albini, a believer in theosophy, astrology, mediumship and numerology. In her autobiography, Lady Dowding acknowledges the immense influence of her mother, writing that her 'most outstanding virtue was compassion'. In line with this Lady Dowding became vegetarian at a young age and felt that the 'spiritual impulse to devote her life to working for the welfare of animals had already been indicated' and she just needed 'a crystallization of that impulse'[113] This duly came in the mid-fifties, when she became involved in both the anti-vivisection and anti-fur movements and finally in 1959, when she established her own campaign and beauty line, Beauty Without Cruelty (BWC). She then became a member of the Council of the National Anti-Vivisection Society and urged her husband, with his seat in the House of Lords, to speak on the issue of animal experimentation. In the summer of 1957 he spoke out pointing out that although the 1856 Royal Commission on Vivisection had been well-intentioned, many anomalies and weaknesses within this law meant that it required significant modification in order to really protect animals in labs. For example the clause detailing that anaesthetic be used during experiments was unclear, which meant that countless animals had been operated on while only

112 'Conscience in World Farming' in *Environmental Health*, February 1975.

113 *The Psychic Life of Muriel, The Lady Dowding, an autobiography*. (The Theosophical Publishing House, London, 1980) p.153.

mildly sedated, or in some instances paralysed but not anaesthetised, so that the animal felt the terrible pain of the operation but was unable to express its distress.[114]

Like Peter, Dowding was far from the crank that the animal movement was oft imagined to be full of, and he in many ways epitomised the establishment. He had been Air Chief Marshal of the Royal Air Force and the officer commanding RAF Fighter Command during World War II, widely considered instrumental in winning the Battle of Britain. In the biopic of his life, Laurence Olivier played Dowding and when Dowding died in 1970 his ashes were laid to rest in Westminster Abbey. Dowding also campaigned for more humane slaughter legislation, and spoke out against the practice of religious ritual killings. From vivisection, Lady Dowding was drawn to the anti-fur campaigns of the late fifties, a very minor campaign at the time, which only gained mainstream force in the sixties, to peak in the eighties and nineties with celebrity and model-backed pressure-groups such as Lynx, Respect for Animals and PETA. By the mid-eighties, anti-fur campaigns had high-profile fashion photographer David Bailey on board (perhaps influenced by his earlier artistic work with Celia Hammond), and he shot for free a now famous ad showing a catwalk model dragging a blood-soaked fur coat accompanied by the slogan: 'It takes up to 40 dumb animals to make a fur coat. But only one to wear it.' A campaign that was, as Lynx co-founder Mark Glover put it, something that turned the tide, 'as people saw the association between supposed glamour and something unpleasant' and 'instead of fur coats being admired, they quickly became something to be ashamed of'.[115] However, two decades prior, Lady Dowding had come to that realisation during a Spiritualists' Association meeting at the Royal Albert Hall when many of the audience were wearing animal fur; it was that that caused her to begin campaigning on the issue. She noted how 'spiritualists tend to talk a lot about "vibrations" and began to wonder what must have been the vibrations of terror and suffering, emanating from the skins of those animals'.[116]

Lady Dowding and a group of women sympathetic to her causes, both spiritual and anti-fur, joined forces to stage a simulated-fur catwalk show in London. She was then led to question other aspects of the fashion and beauty industry, and was shocked to discover, on watching a documentary on the 'slaughter of the great whales … the greatest warm-blooded creature in

114 Extracts from Lord Dowding's speech to House of Lords on 18 July 1957, taken from *The Psychic Life of Muriel, The Lady Dowding, an autobiography* (The Theosophical Publishing House, London, 1980), pp.153–55.

115 https://www.thirdsector.co.uk/change-makers-lynx/communications/article/1289811

116 *The Psychic Life of Muriel, The Lady Dowding, an autobiography* (The Theosophical Publishing House, London, 1980) p.161.

existence, that harpoons were being shot into these majestic creatures for use in such everyday products as soaps, margarines, cosmetic perfumes'.[117] Prompted by this horror she reasoned that 'since everyone uses soap, we ... felt it our duty to try and obtain a soap which did not contain whale-products and to draw up a list of other products that were not the result of tortured animals' bodies.'[118] What Lady Dowding rapidly learnt as she began to compile this list was that, 'animal ingredients, either from whales or by-products of the slaughter-house, were the bases of nearly all cosmetics of the era'.[119] Animal testing within the beauty industry, like factory farming, had grown exponentially after World War II, and like Compassion, Beauty Without Cruelty was a pioneering movement formed at a time when there was little awareness of the industry's cruelties nor any established movement against it. Lady Dowding, in similar vein to Compassion with their aligned businesses of Direct Foods and The Bran Tub, highlighted this callousness of the mainstream cosmetic industry and offered the consumer practical, cruelty-free alternatives. The Dowdings were kindred spirits to the Roberts, and became good friends, both families aware of the hypocrisy of individuals and institutions, and keen to do something about it. The families and charities joined together on many campaigns, as well as making social visits at home.

Dowding created a popular line of vegan cosmetics in an era when tallow (animal fat) was still used in soaps, when whale oil and crushed beetles (shellac) were used for lipsticks, and when musk from deer and castoreun from beavers were routinely used as fixatives for scents. In addition to not using animals in BWC products the company did not test its products on animals. Testing cosmetics on animals is another practice, like factory farming, which can be traced back to America in 1938 and which then became popular in the UK. Like Compassion, BWC began as a volunteer-led and run campaign. When the Roberts family opened The Bran Tub in the seventies a large section was given over to BWC cosmetics.

Of the BWC movement Peter wrote in *Agscene* 'we have constantly been aware that the paths of our two societies lay parallel, both appreciating the need for providing the alternative to cruelty in what might be termed a policy of "constructive abolition" '.[120] As he had for Compassion, the solicitor Noel Earle Gabrielle played a part in helping BWC achieve a more professional and

117 *The Psychic Life of Muriel, The Lady Dowding, an autobiography* (The Theosophical Publishing House, London, 1980) p.165.

118 *The Psychic Life of Muriel, The Lady Dowding, an autobiography* (The Theosophical Publishing House, London, 1980) p.165

119 *The Psychic Life of Muriel, The Lady Dowding, an autobiography* (The Theosophical Publishing House, London, 1980) p.165

120 Peter Roberts, *Agscene*, 1968, Jan/Feb, No.2. p.6.

financially viable standing. Just as he had suggested to Anna and Peter that they ought to just do it themselves when no others within the established animal movements of the time were willing to tackle the farm-animal question, he felt that what BWC was doing was 'something which no other organization in the animal welfare movement was doing … [both] denouncing cruelties … [and] offering cruelty-free alternatives'.

Others on the wavelength were activists Lucy Newman, Clare Druce, and her mother Violet Spalding. Newman had founded the National Society for the Abolition of Factory Farming several years before the Roberts founded Compassion. Her society was the first to fight specifically against factory farming, although unlike Compassion it focused only on Britain. Lucy was an indomitable force, just four days before she died at the age of 92 reported to be marching in Scunthorpe against proposed plans for Unigate to build more 'chicken plants'. A 1975 article in the Agriculture section of *The Times* describes Newman's NSAFF as 'more militant' than other groups of the time, an astonishing description in the context of the NSAFF's charter of beliefs, which states that 'no animal should be removed from its mother until the natural weaning period is over and that antibiotics, hormones, arsenic or any other substance "to promote unnatural development" should be forbidden'. The NSAFF's other 'militant' demands include the belief that 'birds should be able to perch, peck, scratch, preen and spread their wings'.[121]

Compassion later absorbed the NSAFF, providing a united, more powerful, front. Clare Druce and Violet Spalding founded Chicken's Lib just a few years after Compassion in the early seventies, the name of the charity capitalising on the growth of second-wave feminism, which had come to be known as women's lib, in the late sixties. Like the Roberts, Claire and Violet had been influenced by Harrison's *Animal Machines*. Compassion and Chicken's Lib worked together often, including on the production of anti-broiler campaign leaflets in the 1980s and on fact sheets against the new practice of farming ostriches in Britain in the early nineties. Ostrich farms were beginning to take off in Britain due to the potential wealth these large African birds could bring in: not only could their meat be sold, but also their 'leather and feathers' – and indeed potentially every part of the bird – could be turned into something saleable: brooches and earrings from claws, ashtray stands from legs and feet. This great financial potential once again overrode any thought of the animals' needs, for example that an ostrich in the wild will naturally 'run for long periods for speeds of 40 mph' but that on UK farms they were to be 'confined to barren paddocks'.[122]

121 'Animal Lovers Keep Their Eyes on Farmers', by Hugh Clayton, *The Times*, London, 17.11.75
122 Clare Druce, *Chicken's Lib* (Bluemoose, Yorkshire, 2013), pp.204–5.

Joanne Bower was another campaigner who worked closely with Peter and Compassion on a number of campaigns; in 1966, one year before Compassion was founded, she had established the Farm and Food Society (FAFS); Peter had been one of its early members. Bower was FAF's honorary secretary for 36 years, and when acknowledged in her obituary as one of the first campaigners for farm animal welfare was said to have abhorred the stress and suffering inherent in intensive production which at the time few people knew about.[123] Like Peter, Bower felt that the important Brambell Committee and consequent reports had not been taken seriously enough and when the committee's report came to be discussed in Parliament Joanne organised a demonstration and smuggled a cage of stuffed and caged hens into the House of Commons, which the 'MP John Ellis used to great effect during the debate'.[124]

It was at White Eagle Lodge that Anna and Peter had also met Jean Le Fevre, who had become Anna's closest friend. Jean had volunteered for Compassion for many years and had also worked on an immediate and practical level, rescuing countless abandoned and abused animals in need and taking them into her home and garden. As a child, to me Jean seemed the most exciting and mythic of Nan's friends. She wore the pale blue robes of the White Eagle Lodge and worked as a minister and later as a leader at its American and Japanese chapters. She kept dozens of rescue dogs at her home in Crowborough, as well as a gaggle of aggressive geese, and spoke fluent Swahili. Later she moved to Texas, where she founded an endangered animal sanctuary, where she cared for wolves. In photos her grey hair is long and braided, and in one of them she stands with a desert tree at one side and a wolf at the other. She frequently invited Anna to visit her, and though Anna would talk of it she never made the trip.

If, like Jean, someone was on the Roberts' 'wavelength' then the friendship would be intense. As Kim Stallwood recalls, he felt 'taken under Anna and Peter's wing' when he began working for Compassion, and outside the office he spent time at Copse House and took trips with the family to London, including to the West End to see Andrew Lloyd Webber's rock opera musical *Jesus Christ Superstar* – one of Anna's favourites, along with *Miss Saigon* and *Les Mis*. Music, the arts and gatherings of like-minded people were important to both Anna and Peter. Anna loved to blare opera loudly (though everyone else's music, she said, 'hurt her ears!'): Maria Callas, Andrea Bocelli, Pavarotti and Elaine Page were favourite singers. Peter on the other hand still liked his eclectic mix of

123 http://www.organicresearchcentre.com/manage/authincludes/article_uploads/Joanne%20Bower%20expanded%20obit%20for%20EFRC%2014%5B1%5D.3.06..pdf

124 http://www.organicresearchcentre.com/manage/authincludes/article_uploads/Joanne%20Bower%20expanded%20obit%20for%20EFRC%2014%5B1%5D.3.06..pdf

pop and rock: Bob Dylan, Don MacLean, Joan Baez, Paper Lace and Puddle of Mudd. And they both enjoyed singers of the schmaltz variety like Anne Murray and Pam Ayres, and Anna loved to sing the saccharine 'I'd Like to Teach the World to Sing (in Perfect Harmony)' as well as Harry Secombe's 'If I Ruled the World'. In all this a theme was apparent: she wanted to be in charge, and she wanted things to be run beautifully. For Peter, however, music was more about rebellion than harmony. Their other early employee, Elaine Scheperel, far from home and alone during the festive season, was invited to Copse House at Christmastime and recalls Anna's 'delicious nut roast ... the wonderful aromas and the very cosy welcoming atmosphere' of the house.

Elaine – four decades on, and living back in the United States where she teaches singing – still recalls Anna and Peter clearly. Her impression of them is that 'they were utterly devoted to their cause, were kind to all people, and had a zest for living that was infectious. They always seemed to light up a room when they entered it.'[125]

With the help of such wonderful friends and colleagues, the Roberts continued to take Compassion's message out into the wider world. Throughout the decade, they visited veterinary colleges, young farmers' clubs, NFU meetings, MAAF (the Ministry of Agriculture, Fisheries and Food, now Defra) representatives, as well as holding stalls at food and ecological festivals and exhibitions.

At NFU meetings, however, although there was often heated disagreement between Peter and the institution's members, there was still, as former Compassion employee Mark Gold recalled, a sense of 'grudging respect' for Peter, who was largely viewed as 'charming and well liked'. It seemed to help that he was an ex-farmer and that he still dressed like one, usually in a tweed jacket, and that he was polite, logical, well-informed and good-looking! The same three adjectives repeatedly come up: 'intelligent', 'humble' and 'handsome.' A number of female friends and colleagues admitted to 'having a small crush on him' and thought that he looked like Richard Burton or at times Marlon Brando. Adding to Peter's unaffected charm was the yellow Triumph Spitfire that he drove, which became known as 'the little yellow bird' and from which he'd blare either Nazareth, the Scottish hard-rock band, or ABBA. Compassion employee the late John Callaghan described Peter more fully as 'modest, courageous, determined, humorous and decent' and felt that 'he did what he did for his family and animals and was not interested in ego.' John recalled, too, that Peter 'was always such good company, always ready with a joke, always happy to share a drink; that although he was often lauded and acclaimed by others for his achievements he would be the first to make a

125 Elaine Scheperel, personal correspondence, 08/09/15.

joke at his own expense' – and, importantly, was 'also quick to point out that Compassion was started by *two* people – Anna and himself.'[126]

Compassion took its message not only to those working within the agricultural system, directly addressing farmers and farm institutions such as the NFU and MAAF, but also to the general public, by attending every trade fair and festival they could. In 1976, at the inaugural Mind Body Spirit festival in West London's Olympia the Roberts, having set up their stall there, saw a man walk across burning hot coals; at the centre of the hall was a peace pyramid; and the hall jangled with New Age music.

The whole family got involved. Alongside the Compassion campaign stall, Direct Foods gave out samples of Sosmix and Sizzleberg. Here they encountered Buddhists, theosophists, spiritualists, Hari Krishnas, Sufi dervishes; other stalls ranged from serious campaigning organisations to expert practitioners in yoga and authors of books on flying saucers. There were healers specialising in colours, rocks, music and meditation. Kim Stallwood recalled in *Growl* that the crowds were hungry for information at this event and that the show went on for five days – an exhilarating, exhausting, disorienting experience.

Stella Polaris described the festival as an 'annual carnival of all the New Age movements' and at the 1978 Mind Body Festival the Lodge too had a stall, at which members intended to link hands 'with all other workers for the Aquarian Age in harmony and peace.'[127] One might expect to be preaching to the converted or aware at a Mind Body Spirit exhibition but, as Stallwood noted, 'although people were familiar', at least in theory, 'with how veal was produced, the model enabled them to see for themselves just how little space the calves had' – and it was a shock.

They learnt, too, that veal is a by-product of the dairy industry which, as with the majority of factory farm methods, was developed in the fifties. The cows are made pregnant over and over again in order to maximize their milk production, and male calves are taken from their mothers at birth because they are not useful for milk production. Some female calves are also turned into veal if not needed for dairy production. The calves are taken from their mothers usually on the day of their birth, sometimes a day or two after. In natural conditions a calf would spend a year or more with its mother, and countless studies have shown that a cow and her calf develop a 'strong maternal bond in as little as five minutes.'[128]

126 Personal correspondence, John Callaghan, email 6.8.17.

127 Back page of *Stella Polaris*, April/May 1978.

128 https://www.peta.org/issues/animals-used-for-food/animals-used-food-factsheets/veal-byproduct-cruel-dairy-industry/

Exhibitions and trade shows were usually held in London or Brighton, and after intense days of campaigning there was a sense of holiday. In London, dinner out and sometimes trips to the theatre; in Brighton, swimming and shopping in the North Laines, the former slum area that by the seventies was a Bohemian enclave. After packing up a show the Roberts would head to the pebbled beachfront. 'Helen wants to go down to the water to paddle,' Peter would say, though Helen hadn't asked. They would have tea and chips on the beach looking out at the fire-ravaged West Pier, fragments of metal and wood from it scattering the shore.

In 1976, at the somewhat different Daily Mail Ideal Home Exhibition, again in Olympia, tragedy hit, shattering the illusion of the dawning of a golden Aquarian age. Gillian remembers sitting at age 16 in the café of the four-acre hall when they were evacuated. There was a loud noise and she saw smoke but didn't know what was going on and assumed it was just a false alarm or test alarm. The crowds filed out of the building to stand at a distance from the front of the hall as reports filtered through of an IRA bomb. By the late seventies some of the sheen of the hopeful liberation and protest movements of the sixties and early seventies had worn off, and some of those that had 'failed' to achieve freedom had turned to violence. The bomb had been set off at the busiest time of the show, a Saturday afternoon. There were no fatalities but 70 people were injured and four lost limbs.

12

AWOL

Throughout this tumultuous period and as Compassion gained momentum and clout, teenage troubles unsettled the idyll of family life at Copse House. Looking back, my mother Gillian's behaviour doesn't seem extreme for the era, but to Anna at the time it seemed so. Anna entered adulthood in fifties suburban England and retained the prudishness of that era and locality throughout her life, despite so many of her later views being more Age of Aquarian and counter-culture aligned. Beyond Compassion and the animals, the sanctity and harmony of the family unit was the most important thing in her life, and any threat to that she took badly, so her daughter's growing up and experimenting seemed perilous to her. While Peter retained a lightness about life, Anna struggled to accept behaviour that didn't fit into her own moral code, and made this crystal clear to anyone transgressing. She was a (tiny and petite) force of nature!

At fifteen, Gillian took up smoking, found herself drawn to romantic poetry à la Rod McKuen, and got herself a boyfriend. She sewed 'Make Love Not War' badges onto her jeans, and she lied about where she was spending her evenings.

Meanwhile Helen had hit adolescence, and Judy had left home, heading west, to attend Exeter University for teacher training. After graduating, she took a teaching job in a nursery school in Tower Hamlets in East London, close to the famous Bangladeshi-British Brick Lane, where she employed some of her farmer's daughter's knowledge and became involved at the ideas-stage of the creation of Stepney City Farm, a hobby farm with animals and vegetable allotments. The aim was to provide East Londoners, in what was then a mostly deprived area, a chance to experience the amenity of the countryside, providing educational, environmental and creative projects for local children. Stepney

City Farm is now a thriving free-range working farm, community centre and rural arts centre for children and adults.

Gillian was still living at home, with Anna, Peter and Helen, attending Portsmouth's Highbury College, and squeezing into her skinny jeans with the aid of a coat-hanger, when she made a life-changing discovery. She had gone to the family GP complaining of back pain. Anna had attended the appointment with her, and was in the room when the doctor exclaimed, 'No wonder she's in pain, she's almost six months pregnant!' Gillian was 17 years old and single, the baby's father, Paul, by then an ex-boyfriend.

Dominic, my elder brother and Anna and Peter's first grandchild, was born on 7 April 1979. He was a long slender baby with curly brown hair and blue eyes. Anna saw him first; a fact she liked to bring up, saying this made him 'hers'. She was over the moon. Dom was the first Roberts grandchild to be brought up vegetarian.

At 19 Gillian met her future husband in the brutalist-inspired breeze-block Mecca nightclub in Portsmouth. Inside, it was all red plastic tables and chairs with a bar that took up the whole of one wall and a large stage with a glitzy eighties tin-foil backdrop. Kevin hailed from Cwmbran in South Wales, an old coal area by then somewhat deprived, famous for being home to the Jammy Dodger biscuit factory and not much else. Kevin had trained at HMS *Collingwood* in Portsmouth, following in his father's and grandfather's footsteps, both of them proud ex-Naval men, then started work on HMS *Aurora* as an electrician, or 'sparky'.

When Gillian and Kevin met Dominic was two, toddling around in brown cords and a polo shirt, forever holding a blue He-Man doll and hugging guests. The third generation of vegetarian Roberts, little Dominic ended up being the catalyst for Kevin to change his diet. Gillian and her family were the first people Kevin had ever met who didn't eat meat, and he later admitted that prior to meeting Gillian his diet had been fairly 'traditional' and that he'd 'never eaten spaghetti, mushrooms or lentils'. One day while Dom was riding high on Kevin's shoulders as they did the weekly supermarket shop, Dom began asking questions. As they passed the line of freezers, he looked over the top of Kevin's head at the stacks of frozen birds and asked 'But Dad, where are the chickens' legs? Where are their feathers? Where has their skin gone?' Kevin quit eating meat the following New Year's Eve as part of his resolution – a resolution which has lasted over three decades.

For Kevin however a change in diet was not the end of his Roberts-influenced activism. It started to seem incongruous to him that as an ethical vegetarian he was a member of the navy. He reasoned that if he was uncomfortable causing the death of animals, how could he be employed by a business that intentionally killed people? As an electrician he knew it would be unlikely he'd ever be involved in direct combat, but he still felt uncomfortable being a cog in this inherently violent machine. He'd signed up in the first place because he loved the sea and because it was what his father and grandfather had done. He applied to leave the navy before his service contract was up but was denied.

He decided to wait out the conflict in the Falklands, afraid he would be thought a coward if he left during those ten weeks of fighting. His ship was never called to the Falklands, though a false report on the news saying that the *Aurora* had set sail caused panic for his mother and Gillian until they heard from him. His best friend's brother died in the conflict.

The Falklands War ended on 14 June 1982, and shortly after Kevin went AWOL. The phone rang at Copse House. It was his commanding officer.

'Is Kevin with you?' asked the officer. 'Yes … he's just here watching Top of the Pops with Dominic,' Gillian replied.

The officer called him in and asked him to justify himself. He explained once more how his moral view had changed since he'd signed up three and a half years previously. The officer was livid. He was aware of Kevin's girlfriend, an unwed mother, who worked in a health store and had 'turned him vegetarian'.

'Fine – but what are you going to do, then, when the bottom falls out of the cornflake packet?' the officer asked, clearly pleased with his choice of metaphor; presumably he felt that Kevin's new ethics could be summed up by what he saw as 'health food' – the (highly-processed) cereal. For going AWOL Kevin was ordered to relinquish one month's salary and was given ten days of naval punishment number nine. (Number one was being hanged, number nine was cleaning the toilets.) As soon as he'd completed the ten days, he left. the Navy.

The following year, 1984, I was born, one week past due date, I was lucky enough to share my birthday with both David Bowie and Elvis! I came out white-blonde and green-eyed and with a pointed pixie ear.

Anna was less strict with Dom and me and the rest of her grandchildren – Amy, Millie, Holly and Luke – when they arrived than she had been with her three daughters, and as she aged her views seemed to become more permissive. To me she was something of an archetype; forever baking, keeping us cosy, always carrying useful, comforting things in her Mary Poppins handbag; floral hankies that smelled of parma violets, mints, homeopathic sugar pills, and Rescue Remedy. Days out with her were trips to the garden centre, creating herb gardens, planting flowerbeds, baking fairy cakes and watching black-and-white films while eating sandwiches and soup. At night she made us Langdale's cinnamon milk and read us Roald Dahl. She tucked us up in our bunk beds and, when she left, kissed our foreheads. She smelled of lavender and always, without fail, said three things before shutting the door: 'Goodnight. God Bless. Sweet Dreams.'

She encouraged creativity in us and liked it that I ended up studying English Literature, something she would have loved to do. Whenever I had a book lying around Copse House, both Nan and Peter were interested in it. (NB: all of us grandchildren called him Peter at his instruction, rather than Granddad, as he had said it made him feel too old to be labelled that way.) So Peter always seemed to know whatever book I was reading, however niche it might be; he seemed to have a deep knowledge of almost every subject and as a child I believed him to be 'the cleverest person in the world,' a notion I never truly lost. I remember that he didn't always say a lot at family gatherings but when he did his sentences were long and eloquent – and everyone listened. He was sometimes intimidating as well, though, and would speak gruffly even to a thin-skinned grandchild like me. An early memory was of me following their Alsatian, Zara, around Copse House, desperately wanting to stroke her, and Peter telling me crossly to 'stop it and leave her be,' at which I burst into floods of tears. Mostly though, he was playful.

An equally pervasive childhood memory was of mid-sixties Peter spinning us grandchildren around the kitchen. He was babysitting me, aged six, and Judy's two daughters – Amy four, and Millie two – at Judy's house, Sunnyside Cottage, while we were out getting supplies for dinner. When they returned they pushed open the front door to find the four of us, hands held, twirling in a circle and singing, with Peter at the centre getting us to spin faster and faster.

Two years on, Judy and Mike, her husband, had a third daughter, Holly, mimicking the Roberts family with a trio of girls, and in the late nineties Helen had a son, Luke, with her then husband Nicholas. We were all, two generations beyond Anna and Peter, brought up vegetarian, with the choice to eat meat outside the house once we were old enough to make that choice but all of us chose to stay vegetarian, and much of the family has since gone vegan.

By the time I was seven years old, Peter was in his late sixties and newly retired from Compassion. Direct Foods had been sold years earlier, and Gillian had taken over the running of The Bran Tub. The inner workings of Compassion and what everyone had achieved rarely came up during family gatherings. Anna and Peter were humble about it, plus of course we were focused only on that moment – a moment that included dance routines, hot tubs, forest walks, demon drop slides, theatre, tractors and rowing boats. We knew the bare bones of both their mythic love story and the mythic Compassion founding story – a man from the ministry knocking on *their* door of all doors! A tale about monks and veal calves came up occasionally, and there was even the odd exciting time we saw Peter and Anna on the TV being given an award by some celebrity. We were thrilled to discover that they knew Joanna – Patsy of *Ab Fab!* – Lumley, a long-term and outspoken patron and supporter of Compassion's work. But the full extent of their achievement wasn't visible to us, nor was the extent of their rebellion. Their views, by the time we grandkids were able to understand them in the nineties, seemed more or less mainstream; we were a clearly in a minority being vegetarian, still regularly asked by school friends where we got our protein from, and (annoyingly) 'Would you eat a chicken if you were stranded on a desert island and starving?' but it seemed clear that even if most people still wanted to eat animals they at least wanted them to have 'a nice life' in a field first, rather than being factory-farmed. This was easy to agree on, it seemed. What we failed to appreciate was that it was Grandma and Peter who had played such a large part in making these views – views that had once seemed radical and crankish – now seem mainstream, decent and even normal.

13

A DIRTY WORD

> It's crazy that the idea of animal rights seems crazy to anyone. We live in a world where it's conventional to treat an animal like a hunk of wood, and extreme to treat an animal like an animal.
>
> Jonathan Safran Foer, *Eating Animals*

In the eighties, Anna and Peter moved Compassion out of Lyndum House to an office space above The Bran Tub (later they moved again into even bigger offices on Charles Street, just up the road from the shop). The campaign had grown and Compassion had begun to gain international clout with partners not only in Europe but worldwide; as the name suggests, this had always been the plan. By the mid-eighties Compassion's affiliated charities included: Animal Liberation in Sydney, Australia; the Anti-Cruelty Society in Palmerston North, New Zealand; the Farm Animal Concerns Trust in Chicago, USA; the Farm Animal Reform Movement in Washington DC, USA; and the Canadian Federation for the Protection of Animals. The Roberts vision had always been an international one, encouraging the various worldwide animal groups to abandon their traditional 'credo of splendid isolation', as Peter put it, and join forces.

More people were, it seemed, finding their wavelength. The late John Callaghan, who had worked for Compassion as international development director with animal welfare groups in Central and Eastern Europe, recalled how

Peter was always so interested in his work and so supportive. It was great working with him because he and Anna were the ones who started it all, and it made you feel so proud that you were helping to fulfil their dreams of improving the lives of millions of farm animals. Peter really supported the international work as he wanted us to really become a worldwide organization.'

Compassion now had a growing membership and prestige, yet its campaigns remained largely unchanged and despite the growth the office was still, until the early nineties, a barebones team of ten paid staff. The Roberts and their team of workers and volunteers campaigned for a ban on the battery cages and intensive broiler units for chickens and laying hens; for a ban on the dry sow-stall, for a ban on the production of foie gras, for a ban on veal-calf crates; and for a full ban of the live export trade. Alongside this, they opposed lesser-known worldwide practices such as the growth of frog farms in India, where poor workers were paid a pittance to chop the legs off live frogs to feed to Western markets; on a ban on the grinding of sheep's teeth in Australia (a ban that they achieved) and a ban on the Australian practice of mulesing (an extremely painful operation, slicing off the folds of a sheep's skin to create stretched scar tissue, reducing the risk of flystrike), on curbing the use of growth hormones and antibiotics on farms across the EEC, and against the ecologically disastrous British practice of straw-burning.

Yet whilst these campaigns, both major and minor, have since 1967 met with measures of success, the world's ever-growing population and resultant factory food system's growth mean that Compassion's struggle is still constant, still great.

The 1980s however did prove to be a landmark decade for animal welfare, and in 1983 Compassion began a course of action that was to prove monumental in terms of the charity's political and public power – and one that, most significantly, was to reduce the suffering of millions of farm animals. It was one of the celebrated, seemingly sensationalist, family tales that we knew from an early age. To a child's mind it was riveting: our grandfather had taken monks to court! The details were hazy, but even without the ins and outs we knew the story had a whiff of excitement and rebellion about it.

The Catholic monks of Storrington Priory had been resident in the village of Pulborough in West Sussex for over a century when the scandal that was to both shame their order and change Britain's agricultural law hit them. Peter's attention was drawn to the Norbertine Canons of Storrington and their veal farm, where Compassion found 650 calves kept in crates and chained by the neck, confined on the usual slatted floor system and denied the roughage in their diet that they crave and that their bodies require. Following a six-week life of undeniable suffering the calves would be loaded into trucks and sent on an arduous and stressful journey to the Continent, where they were then subjected to further deprivations until their slaughter.

In the eighties, when Peter and Compassion challenged Storrington Priory, mainstream veal farming practices included removing the calf within a day or two from its mother, which is stressful for both calf and cow (the cow will often cry out for her calf for hours or even days after it has been taken). Other palpable signs of stress in the cow include raised heart rate, increased visible eye-white and reduced rumination. As far as the calf is concerned, when he is taken away he loses key health benefits; his mother's licking should have been stimulating and assisting his breathing, promoting his circulation and encouraging normal, healthy urination and defecation. Veal calves kept in such crates regularly suffer from chronic diarrhoea, and without colostrum, the highly nutritious calf's first milk, they suffer poor immunity so are much more prone to infection and disease. A crated calf spends his entire life indoors, experiences prolonged sensory, social, and exploratory deprivation, and is fed a synthetic formula, intentionally low in iron and other required nutrients; anaemia is desirable from a factory-farmer's point of view as it makes the animals flesh more tender (known as 'white veal') for the consumer. The wobbly walk out of the veal crate and into the slaughterer's truck marks the second time that the animal might ever see daylight.

John Webster, Professor of Animal Husbandry and Head of the Veterinary School at the University of Bristol and one of the UK's leading experts on dairy cattle described the veal crate system as it was in the UK until 1990 as 'one of the most bizarre and, in my opinion, unequivocally cruel forms of livestock production'.[129] Under a decade into the rise of veal in 1960 conservative British newspaper the *Daily Mail* criticised the veal crate, stating that the system was 'so unnatural as to outrage ordinary human feeling'.[130] Peter called the crates 'premature coffins.'

The historical, mainstream Catholic view and treatment of animals has largely been poor, with some wonderfully notable exceptions such as that of Saint Francis of Assisi, who died on 3 October 1226, at the age of 44, and was canonised just two years later, in 1228. He is now recognised as the patron saint of animals and of ecologists, honouring his boundless love for animals and nature. World Animal Day is now celebrated each year on 4 October, his feast day.

Francis's life followed the trajectory from sinner to saint. He had reportedly been a medieval playboy with ambitions of becoming a knight, and after a night of heavy drinking he broke the Assisi town curfew to end up in prison. That night he received visions which he believed to be the direct word of God. The playboy antics ceased, he kissed lepers in the street and he spent his hours

129 Internal CIWF document, CIWF Founding Ethos by Carol McKenna.

130 Internal CIWF document, CIWF Founding Ethos by Carol McKenna.

meditating on mountains and in quiet churches. He retained his charisma, however, and was soon drawing thousands of followers, preaching in up to five villages a day, teaching a new kind of emotional and personal Christian religion that everyday people, not just those educated enough to read the scriptures, could understand.

While some regarded Francis as a madman or a fool, many Christians view him as one of the greatest examples of 'how to live the Christian ideal since Jesus Christ himself'.[131] Francis was supposedly the first person to receive Christ's stigmata and, like Christ, was a revolutionary figure. One of his key tenets was respect and compassion for animals and the natural world, and he took this so far as to preach directly to animals.

In every October edition of *Agscene*, Peter would draw the readers' attention to St Francis's Day as a good time to examine our relationship with animals, posing the questions, 'Do we accept our responsibility for their welfare?' and 'Is the church itself sufficiently vigilant?' He pointed to 'the shame of the church, that with few notable exceptions, has never risen to the defence of factory farm animals'.[132] Eight years after penning this article Peter found that not only was the church as an institution failing to rise to the defence of those animals but that some members of it were actually profiting from their suffering.

The Storrington Priory followed the rule of Saint Augustine of Hippo, one of the forefathers of Christian Catholic doctrine. Augustine's 5th-century, powerfully titled book, *City of God*, inspired by the infamous Visigoth sack of Rome, covered the big issues of the time: the conflict between the existence of free will and divine omniscience; the implications of original sin; and the meaning of Grace and Salvation – Grace as God's love and mercy, Salvation as God's redemption. Augustine is also credited as the patron saint of brewers, printers and theologians.

Key to the Norbertine Canon followed by the Storrington monks was a focus on austerity. Undoubtedly, the Priory's keeping of animals in confined crates was a far cry from the preachings of St Francis, and their profit from cruelty was at odds with the abstemious and mercy-focused nature of the Augustinian order.

Peter reported in *Agscene* that the 'Priory was the first independent House of the White Canons to be established in England since the Reformation. It is significant that it is dedicated to the Mother of Jesus to bring many souls to Jesus through Mary's loving intercession.' There is precious little respect for motherhood in veal farming, where calves are taken from their bellowing mothers at a few days of age to be chained immobile on a slatted floor for life,

131 https://www.biography.com/people/st-francis-of-assisi-21152679

132 Peter Roberts, *Agscene*, October 1975, p.1.

and then shipped to their deaths in a foreign slaughterhouse.[133] 'In legal terms the monks' farm was perfectly lawful in Britain at this time; in fact the veal unit had been built with a government grant.' [134] But Peter knew that the keeping of motherless calves in barren crates by an ascetic religious group would ignite public outrage, and he hoped that this outrage would spark real change.

In 1983 Compassion employee Lesley Turpin led a protest outside Storrington Church along with supporters from aligned groups including Animal Aid, and supporters from the Catholic Study Circle for Animal Welfare The protestors handed out leaflets to the churchgoers detailing the treatment of calves in their own parish, and waved banners saying 'religious killers'. They accused the monks of making profit out of animal suffering. Some parishioners were supportive, giving donations to Compassion as they left the church, and others joined the protest. Local news reported:

> Josephine Potter, 63, of Chatsworth Road, Brighton, who came along with her family, stated that she 'was married in this church and that her children were baptised here' and as such she felt 'horrified to read what was happening, and that it was a 'cruel and unnecessary way of keeping animals.' She concluded that 'as a practising Catholic she felt she had to come along to protest'.

Compassion reported that they had 'twice visited the farm but were not allowed access to the calves' but that they had been 'informed that the calves live, chained individually by the neck, and lie on wooden slats without straw bedding'.[135] They wrote that on

> one of our visits there was a large cattle-truck from St. Malo, France, and it is evident that there is also involvement in the lucrative live export trade to the Continent. Questions are being asked about the monks' life style allegedly from the proceeds of factory farming and the live export trade'.[136]

In response to Compassion's protest the monks issued a leaflet stating 'We take a pride in our animals while they are in our care.' Father Kirkham of the Priory told the press that 'on moral grounds he could not see any objection

133 http://www.all-creatures.org/fol/art-20120917-04.html

134 http://www.all-creatures.org/fol/art-20120917-04.html

135 http://www.all-creatures.org/fol/art-20120917-04.html

136 http://www.all-creatures.org/fol/art-20120917-04.html

whatsoever' and he quoted the Bible in his statement:, 'You only have to look in the scriptures ... there is talk of killing the fatted calf for festivals.' Father George Joy of the Priory said, 'the farm conforms to Ministry of Agriculture's regulations and has a high level of hygiene'.[137]

Peter responded to Father's Kirkham and Father Joy with a statement that read 'the monks in Storrington Friary did not keep a fatted calf but rather 650 calves which they rear for white veal in narrow and barren crates'.[138] He continued

> [i]t may be argued that God blessed the calves giving them four legs and the good earth to walk upon, but the wooden crates that imprison them now do not permit the cows to walk or exercise, nor even so much as turn round. And furthermore on their all-milk diet the only fibre they find is hair plucked from their own bodies, or splinters from their wooden crates.

He closed with reference to the mainstream Catholic belief 'that the Scriptures allow them to do these things and that anyway, animals cannot really suffer'.[139] Many Catholic authorities at the time believed animals to be soulless, a belief that has historically justified Christians' use and abuse of them.

This is a belief challenged by some within the faith, most notably in recent years by the Pope at the time of writing, Francis.[140] The Compassion–Storrington protest made the headlines, and Compassion decided to capitalise on this by taking the monks to court. Media interest was intense, the local Catholics started going elsewhere to mass, and the then Pope, John Paul II, received hundreds of letters of protest. It was a PR disaster for the monks, for the owners of the priory, and for the Catholic Church at large.

A telephone conversation tape-recorded by Peter between him and Mrs Ruggiero, one of the farm managers at Storrington, gives some idea of the tense atmosphere at the time.

> R: We've had a notice from one of your supporting groups, the ALF [Animal Liberation Front].
>
> P: No, it's not one of our supporting groups. We don't have anything to do with them.

137 https://www.all-creatures.org/fol/art-theother.html

138 https://www.all-creatures.org/fol/art-theother.html

139 https://www.all-creatures.org/fol/art-theother.html

140 https://www.csmonitor.com/The-Culture/Religion/2014/1212/Did-the-pope-just-say-that-animals-have-souls

R: Well, I thought you might say that. You see, we were sent a postcard in a sealed envelope about liberating animals, people on the front wearing balaclava-type helmets … Terrorists, you know, with the headgear. I thought I'd let you know that I've informed the police and given them photocopies of the correspondence from them.

P: Mmmmmm. Quite right; I think that's what you should do.

R: We feel, though, that it's a result of the ad you had.

P: I don't understand, the ad? … You mean what we published in *Agscene*?

R: Yes. It could be as a result of that. We've had a lot of post of that nature since, you see. We've had a lot of … publicity … for want of a better word.

P: Oh, I expect you have. But I must correct you when you say that it's a supporting group of CIWF, because we have no affiliations, no contacts with the ALF. I know about them; we see a lot of press cuttings about them. We have been asked to support them, and we have declined. In fact, there was an incident in Bristol when they threatened to use our name because we were not militant enough. They wanted to involve our name in their activities, and I reported the threat to the police. So, you see, we don't have anything to do with them.

R: I thought I'd let you know anyway.

P: Hmm, well, okay, I'm grateful. (Peter's tone is cynical, clearly tired of Ruggiero's repetition here, though he remains polite.)

R: All right. I thought I'd let you know. But on page 4 of your Ag mag you suggest a day, and that the following day be a public awareness day, by announcing it to the press, having vigils etc. Well, I'm just … you understand?

P: I understand that you're trying to make a veiled threat to me of legal action or the like.

R: I'm just advising you what we've received.

P: Well, you've done that. Goodbye. (The phone is hung up.)

Those opposed to the animal welfare movement tend to be categorised in one of two polarised stereotypes: the little old lady, or the balaclava-clad, militant, misanthropic extremist. Basically it's diminishment or demonisation, and Compassion over its 50 years has been victim of both. In recent years and especially since 9/11, the latter narrative seems to have triumphed and it is a rhetoric that has led to real and worrying legislative changes – mainly in the USA, but also elsewhere in the world. Writer and civil liberties advocate Will Potter in his exposé, *Green is the New Red*, highlights similarities between American Senator McCarthy's communist 'red scare' of the forties and fifties, and what Potter calls the new 'green scare'.

Now, just as in the fifties, FBI intimidation is common. Potter was prompted to write by an FBI visit to his home; two agents had threatened his reputation as a journalist at the *Chicago Tribune* – and had suggested, too, that his girlfriend's PhD funding would be cut if he didn't cooperate with them. They had also threatened to put him on a 'domestic terrorist' list. Potter explained that this visit had apparently been prompted by his distribution of some leaflets in a residential neighbourhood, the leaflets containing information about a local animal experimentation lab. That was the extent of his 'crime'.[141]

In America today, detention for environmental and animal rights protestors who commit non-violent crimes such as property damage and animal liberation are routinely, disproportionately lengthy, and the people concerned are regularly kept in maximum security prisons alongside violent criminals. Visiting hours are severely restricted and solitary confinement is common.

To date there are no records of death or injury to humans because of the actions of the Animal Liberation movement anywhere in the world, (though members of the movement itself have been killed while campaigning), yet in recent years, as concern has grown over the threats of climate change and the role played by farmed animals in that, specific legislative changes have been developed and passed which target animal rights and environmental protestors under the guise of protecting the public from 'eco-terrorists'. Such laws have often been passed by Congress, as Potter noted, with less than 1 per cent of representatives present.[142] In many American states so-called 'ag-gag' laws prevail, which mean that a citizen can be jailed for taking a picture of animal mistreatment as they merely pass by – not break into – a farm or abattoir. Ag-gag laws currently exist across Kansas, Utah, Montana and Indiana as well as in the more liberal states of New York and California.

In 2015 in Canada, so often seen as America's more progressive neighbour,

141 http://willpotter.com/bio/

142 https://www.ted.com/talks/will_potter_the_shocking_move_to_criminalize_non_violent_protest#t-237011

a woman tried to give water to pigs in a tractor-trailer stopped at an intersection on its way to Freemans Pork processing facility. She was arrested. The court eventually found her not guilty, but until that point she had faced the possibility of ten years in prison.[143] As Potter said in a 2011 TED Talk, 'the FBI's training documents on this so-called "eco-terrorism" are not about violence; they're about public relations.'

The worry for the environmental movement in Britain and Europe is that where America leads we will follow, as was the case with factory farming in post-war Europe.

On a second recorded tape there is a conversation between Peter and Mary J. Barrington, the barrister who worked on the Storrington case for Compassion, in which Peter practices his opening court statement. It reads: 'The level of deprivation inflicted by the monks of Storrington Priory contravenes the EU convention, the cattle code, the Brambell Report, and the Animal Protection Acts of 1911 and 1958.'

Compassion took the priory to court under nine counts of cruelty. After weeks of proceedings, the judge ruled that the Priory's intensive unit had not caused 'unnecessary suffering', as the conditions were standard for the era (which just about says it all!)

Compassion was ordered to pay £12,000 costs, a huge sum for what was then still a small charity. As Peter had expected and hoped, however, the case proved to be a disaster for the priory, and in October 1984 – just a year after the start of the campaign – *Agscene* reported,

> MONKS' VEAL FARM FOR SALE: The priory at Storrington has announced that it is to sell the veal calf rearing unit and farm for 'economic reasons'.

The article went on to 'thank everyone who campaigned towards this successful conclusion, and all who have supported us in the Court Action and the High Court Appeal'. Making Compassion's determination clear, the piece concluded that the building had 'been put in the hands of the estate agents as an intensive calf unit with about 750 calf stalls' and that anyone 'buying it and intending to carry on its present usage will need to be advised of the level of public hostility against it'.

Because of the publicity generated by the priory case, the people of the UK began boycotting white veal. With the evidence of declining sales and public

143 https://www.theguardian.com/world/2015/nov/30/canada-woman-10-years-prison-for-giving-pigs-water

hostility toward the veal crate system, Compassion's campaigns director Carol McKenna spoke to the National Farmers Union and urged them to enforce a *total* ban on the system. She pointed out that there were only about four or five veal farms left in the whole country, and suggested that the NFU would be doing themselves a favour by following through with a ban, because the veal crate system was shaming the very institution of the NFU. If they failed to act, Carol told them, then veal would soon become such a dirty word that the product would never recover.

The NFU's policy, however, seemed to be that if even one farmer wanted to farm using the veal crate system then the NFU would support said system. However, in 1986, a few months after the court case had been heard and three years on from the start of Compassion's campaign against Storrington, MAFF insisted that Peter attend a conference at the National Agricultural Centre in Stoneleigh where then junior minister, Donald Thompson, was due to speak. Having no idea why he was being invited to the conference, Peter suggested Carol or Joyce in his place, but MAFF insisted that Peter attend.

Donald Thompson was a grammar school Yorkshire man, a farm owner and a Conservative MP, a solid man, sanguine in outlook; his mother was a weaver who'd left school at twelve, and his father owned a family butchers which specialised in black pudding. He sounded as though he'd stepped out of the Canterbury Tales. To Peter's utter amazement, Thompson announced that the UK government would ban the keeping of calves in veal crates. The ban became law in 1990. Although the move was by then more academic than anything else, it demonstrated the power of public opinion and ensured that veal could never make a comeback should it recover from the Storrington PR disaster. It was Compassion's first major victory, and continued pressure resulted in legislation to ban veal crates across Europe by 2007. Sadly the veal crate system is still used elsewhere in the world, most extensively in America.

That night on the drive home from Warwickshire, Peter, Joyce and Carol pulled over at a roadside garage to buy a bottle of champagne, which they drunk from cups in a layby; as Peter put it, 'We may have lost the battle but we have *certainly* won the war!'

14

ATHENA

Compassion was now well into its second decade, and Anna's, Peter's and the charity's reach and interests were expanding, and so alongside the on-the-ground campaigning Compassion's academic wing began to grow. An L-shaped room known as The Den at Copse House was where Peter would research, read and write; there I would spy him peering at massive yellow-papered tomes through his magnifying glass. The view from The Den on one side was the tree- and tulip-filled garden, and from the other window the grey stone patio leading into Great Woods. The air in the room was pleasantly musty, and laid on his desk were always stacks of notes, a massive Oxford dictionary and a selection of daunting (to me, anyhow) hard-backed books. The desk was covered in writings in his spidery scrawl, ranging from French, Latin and Modern Greek language lessons to particularly interesting passages copied out from *The Origin of The Species*, Richard Dawkins' *The Selfish Gene*, and E.L. Grant Watson's *Man and His Universe*. The Den's ornaments, art and books summed up Peter's eclectic taste very well, the objects and scribblings in the margins of his favourite books all clues to his personality, and his numerous trophies testament to his achievements. (Anna and Peter were always at odds about books. Anna would open one so delicately she'd barely crack the spine; she would always use bookmarks and would never dream of using a pen in a book. Peter, on the other hand, folded interesting pages, underlined important points in red biro and starred his favourite passages, and on the inner covers wrote extensive notes apparently in dialogue with the book.)

A trio of hand-made matt black and orange plates showing classic Greek myths hung above his desk. They had been picked up at a pottery during a family holiday in Cyprus, and showed the Olympian god of the sea, Poseidon, his sea-goddess wife Amphitrite, and Demeter, goddess of the harvest, holding her long-stemmed grains of wheat and corn. On Peter's desk was a framed photograph of his beloved brother Frank, and on the whitewashed wall a reproduction of Salvador Dali's 'Christ of St. John of the Cross' and a small nautical print of a lone ship on a choppy sea. On the exposed stone of a disused fireplace hung a crest of two long faux-medieval swords, crossed, on a red and gold plaque. A book-lined wall, seven shelves to the ceiling, held the rest of the library; Peter's breadth of taste meant that his collection ranged through the various disciplines as thoroughly as any public library.

Anna's books were there as well. Her taste was expansive, too, but in terms of fiction she tended to favour romantic and above all *wholesome* writing. 'The only four-letter word in my books is love', said Catherine Cookson, and Anna approved, reading all of that massive oeuvre as well as her autobiography. Beyond Cookson, Anna's favourite novel was Rohinton Mistry's seminal masterpiece *A Fine Balance* (she gave me a copy of it, saying that the descriptions in it took her right back to India which she and Peter visited in the late eighties and early nineties). The rest of Anna's collection included books about ducks, horses, alternative healthcare, religion and gurus.

Peter, on the other hand, loved the classics: Shakespeare, Dickens, William Blake with his mystic visions – and especially Vita Sackville-West's lengthy and lyrical poem 'The Land', which recorded a year in the life of the soil, trees, plants and animals of the Kent countryside. Peter also read theology, both mainstream and mystic; the shelves of The Den bowed under the weight of the huge tomes of Russian occultist H.P. Blavatsky's *The Secret Doctrine*, plus the Vedantic Hindu *Isha Upanishad*, an illustrated Holy Bible, *Jesus in India*, *The Mystery Religions of the World*, *A Rosicrucian Q and A* and many, many more. The classics shelf was the fullest; he was obsessed with the culture and myths of ancient Egypt, Rome and Greece, and took many holidays to the Greek mainland and islands, filling album after album with, as his youngest daughter Helen put it, 'pictures of pillars and posts'. In 1986 he extended his knowledge further, taking an Open University course on Greek History, focused on 4th-century BCE Athens, an era of 'peak achievement in art, literature and philosophy, yet socially and politically … one of dramatic change and conflict'.

That same year he established an education and research trust to complement Compassion's practical campaign work, and named it the Athena Trust after the Greek goddess of reason and wisdom. Its aims were

> to further the educational work of CIWF … to promote and advance the education of the general public in the science of animal behaviour, in the field of knowledge known as ecology,

> including man's activities and the relationship of those activities to the biosphere of the earth.

The Athena Trust was 'dedicated to the promotion of harmony between people and the natural world'.[144] Athena was said to have sprung full-grown and armour-clad from the forehead of Zeus, her father. For some classicists this marks the beginning of a patriarchal culture because the goddess is born from the head of a *male* god. Gender politics aside, it's clear that Athena's strength lies in her faculty of reason, so the Athena Trust in itself constituted a challenge to those who believed that animal welfarists were simply sentimentalists.

The Trust researched and campaigned on a diverse range of issues, from the indiscriminate use of BST (bovine somatropin, a genetically engineered growth hormone that's injected into dairy cows), to the ethics of bio-science and cloning, to religious slaughter practices and issues of ecological degradation. The Athena Trust joined other animal welfare and environmental groups as well as concerned consumer groups and a range of scholars from the world's major faiths. They partnered with a diverse mix of societies including Animal Aid, the Vegetarian Society, Greenpeace, the Maternity Alliance, the London Food Commission, the National Federation of Women's Institutes, the European Centre for Islam, the Anglican Society for Animals and more.

One of their first major projects was to reach out to various religious authorities, and in their first year The Athena Trust commissioned esteemed Muslim scholar Al-Hafiz B.A. Masri to write *Animals in Islam*. The title Al-Hafiz denotes someone who has memorised the Qur'an in its entirety and in 1964 Masri was the first Sunni Muslim to be appointed as the Imam of the Shah Jahan Mosque in Woking, then the European Centre for Islam. In his preface to *Animals in Islam* Masri explains that 'Quite a few … friends have been surprised to learn that I have chosen animals as a subject to write on from the Islamic point of view … they feel that I should be more concerned with the multifarious other problems Muslims are facing these days'.[145] He goes on to justify his choice with the rationale that 'life on earth cannot be disentangled for the amelioration of one species at the expense of the other … that like human beings, animals too have a sense of individuality' and he asks 'how right is it to deny these creatures of God their natural instincts so that we may eat the end product?' *Animals in Islam* included chapters on humanity's dominion over animals, the degrees of animal consciousness, the ethics of

144 *The Bio-Revolution: Cornucopia or Pandora's Box?* eds. Peter Wheale and Ruth McNally (Pluto Press, 1990), Preface, p.1.

145 Al-Hafiz B.A. Masri *Animals in Islam* Preface (Athena Trust, Hampshire).

vegetarianism, the issue of Halal slaughter and the importance of respecting the balance of nature.

On the issue of factory farming Masri wrote

> Our Holy Prophet's overwhelming concern for animal rights and their general welfare would certainly have condemned (*la'ana*) those who practise such methods [factory farming], in the same way as he condemned similar other cruelties in his days. He would have declared that there is no grace or blessing (*brakah*) – neither in the consumption of such food nor in the profits from such trades.[146]

He also wrote

> If animals have been subjected to cruelties in their breeding, transport, slaughter, or in their general welfare, meat from them is considered impure and unlawful to eat (Haram).

The issue of halal meat is a contentious one; however it was not an issue that Masri, as a prominent and well-respected Islamic leader, shied away from. The Arabic word *halal* means 'permissible', and the rules of slaughter are based on Islamic law. To be halal

> the animal has to be alive and healthy, a Muslim has to perform the slaughter in the appropriate ritual manner, and the animal's throat must be cut by a sharp knife severing the carotid artery, jugular vein and windpipe in a single swipe. Blood must be drained out of the carcass.[147]

The halal justification for this draining of blood is a medical one, as blood is a carrier of disease; this too explains the Muslim prohibition on eating pork; pigs being genetically similar to humans means that it is easier for pathogens to be passed from pig to human than from other 'food animals'. Masri took a strong stance against the inhumane aspects of modern, mainstream halal methods. The tenet of halal law stating that the animal has to be alive has often

146 http://www.call-to-monotheism.com/animals_in_islam__by_al_hafiz_b_a__masri

147 https://www.theguardian.com/lifeandstyle/2014/may/08/what-does-halal-method-animal-slaughter-involve accessed 26.5.17

been interpreted to mean that farm animals being slaughtered should not be stunned, or only stunned with a very low voltage, in case this stunning kills rather than knocks the animal unconscious. Masri explains that,

> The main counsel of Islam in the slaughter of animals for food is to do it in the least painful manner. All the Islamic laws on the treatment of animals, including the method of slaughter, are based in all conscience on 'the spirit' of compassion, fellow-feeling and benevolence. Failure to stun animals before slaughter causes them pain and suffering. Muslims should give serious thought to whether this is cruelty (*al-muthiah*). If so, then surely the meat from them is unlawful (*haram*), or at least, undesirable to eat (*makruh*).'[148]

Masri cited contemporary research from Al-Azhar University in Cairo to reinforce his opinion on stunning and proper halal slaughter practices. He writes that the university had appointed a special committee to decide whether the meat of animals slaughtered after stunning was lawful; the committee consisted of representatives of the four acknowledged schools of thought in Islam, i.e. *Shafii*, *Hanafi*, *Maliki* and *Hanbali*. The unanimous verdict (*fatwa*) of the committee was:

> Muslim countries, by approving the modern method of slaughtering, have no religious objection in their way. This is lawful as long as the new means are 'shar' (*Ahadd*) and clean, and do 'cause bleeding' (*Museelah al-damm*). If new means of slaughtering are more quick and sharp, their employment is a more desirable thing. It comes under the saying of the Prophet(s) 'God has ordered us to be kind to everything' (*Inna'l-laha Kataba-'l-ihsan 'ala kulle Shay'in*). Masri adds more scholarly evidence to support his views with the conclusion that, 'To crown all verdicts (*Fatwa*), here is the 'Recommendation' of a pre-eminent Muslim organization of this century – The Muslim World League (Rabitat al-Alam al-Islami). It was founded in Makkah al-Mukarramah in 1962 with 55 Muslim theologians (*Ulama'a*), scientists and leaders on its Constituent Council from all over the world. MWL is a member of the United Nations, UNESCO and the UNICEF. In January 1986 it held

148 http://www.call-to-monotheism.com/animals_in_islam__by_al_hafiz_b_a__masri

> a joint meeting with the World Health Organization (WHO) and made the following 'Recommendation' about pre-slaughter stunning Pre-slaughter stunning by electric shock, if proven to lessen the animal's suffering, is lawful.[149]

However, the majority of halal abattoirs in the mid-eighties, when the Athena Trust published *Animals in Islam*, both in the UK and Europe were not following the guidelines given by Masri, the University of Cairo and the MWL for a humane and Muslim-appropriate slaughter. In a mid-1980 edition of *Agscene* Peter noted the growth of the halal meat packers, and exposed the myth that 'Halal is supplying only the Moslem market'. Peter pointed out that France's Intermarché supplied to all France, with its 850-plus supermarkets, despite the fact that the Muslim population at the time was mainly in Paris and Marseille. In the UK. too Peter noted it could not be denied that with '13,000,000 animals being ritually killed in the UK every year, it is no longer a religious concession to an ethnic minority' and that the 'law should take account of the way it is being abused, and reform the slaughterhouse legislation to ensure one law for all'.[150]

As Peter suggested, halal growth across Europe seemed to be more about slaughterhouse profits than religious belief, and so in 1986, backed by Masri and other prominent leaders in the Muslim community, Peter led Compassion on an anti-halal protest march. The peaceful and Muslim-supported march was, however, usurped by the right-wing, extremist British National Party (founded in 1982), infamous in Britain for their xenophobic, racist and Islamophobic views. Peter was devastated at this exploitation of Compassion's anti-cruelty message, one which had the backing of many Muslims; asserting 'never again', he refused to organise any more anti-halal demonstrations during his lifetime.

Beyond working with the Muslim community and Muslim leaders, the Athena Trust worked closely with various leaders from the Church of England. One such ally was British Anglican priest, professor and author Andrew Linzey, is a member of the faculty of Theology at Oxford University. He held the world's first ever academic post in Ethics, Theology and Animal Welfare at Blackfriars Priory, a Dominican priory in Oxford with close ties to the prestigious university. (The Dominican order was founded in 13th-century Toulouse by the Spaniard Santo Domingo).

Linzey wrote several books on the theme of animal suffering in relation to Christian doctrine, including *Animal Rights: A Christian Perspective* (1976),

149 http://www.call-to-monotheism.com/animals_in_islam__by_al_hafiz_b_a__masri
150 Peter Roberts, *Agscene*, December 1986, no page.

Christianity and the Rights of Animals (1987) and *Animal Theology* (1994). He had collaborated with Peter since the very start of CIWF and, having met in person at the Cambridge Animal Rights Symposium, they campaigned together at the Oxford Animal Fair protesting against live exports from Dover; at the time Linzey was a curate in Dover. Compassion worked alongside the Anglican Society for the Welfare of Animals (ASWA), and an illuminating letter from Mrs Y. Coaks, editor of that society's newsletter, gives some insight into Compassion's engagement with such religious groups:

> Dear Mr Roberts,
>
> I am sending you a book by Victoria Lidiard (whom I suspect you remember – she was very active in the export of live animals campaign) because I think it is one of the few books which puts the animal question right at the centre of Christian living.
>
> The question has been asked more than once: 'Is Christianity a cruel religion?' It came as a small shock to me when I first heard it. How could anybody who has read the gospels of St. Paul in 1 Corinthians chapter 13—ask it? But then I remembered the chequered history of the Church which has at times fallen into all the sins which beset humanity – including cruelty … The questioner however was not thinking of the historical church, but the Church today, and in the context of the Christian relationship with non-human creatures. What had Christians to say about the suffering of non-human creatures which is inflicted daily upon millions of animals in every conceivable way wherever men can derive some pleasure, profit, or believe they can benefit from it? What sort of God do they worship who appears unconcerned about the suffering of animals inflicted on them by humans?'

Mrs Coaks' questions, asked from her Anglican perspective, were what Peter posed to the believers and leaders of all faiths and denominations; he felt strongly that the faithful demographic was one that Compassion must call to account with respect to animals, ethics and the environment, and from that foundation he appealed to church, mosque, synagogue and temple. But Masri, Linzey and Coaks were anomalies within the Judaeo-Christian communities, and Peter frequently lamented that very few spokespeople from the major faiths were speaking out on behalf of animals.

There was at least Joan Watson of the Christian Consultative Council for the Welfare of Animals, who wrote in *Agscene*

> that though several eminent church leaders have spoken out against factory farming, animal experimentation and killing for fun or adornment ... sadly, the established church is silent still.[151]

Watson called the church to wake 'from its seeming apathy towards the suffering of God's defenceless creatures'.[152]

Peter asked, 'How is it that the Church is silent?' especially when there were then (the early eighties) no less than 24 bishops in the House of Lords, and he concluded that 'their silence is an insult to their calling'.[153]

The Athena Trust worked alongside religious groups in terms of addressing the suffering of individuals animals, as well as in relation to species as a whole, and as science progressed Peter spoke out more and more on the issues surrounding bio-ethics. In 1988, he asked if the church could be 'relied upon to influence control over the use of genetic engineering' and noted that although the Archbishop of York had

> recently called for strict international rules to govern the use of genetically engineered organisms that apart from this initiative and those of other notable individuals, such as the revered Dr Andrew Linzey, the Church had shown a marked reluctance to get involved in the ethics of science.

In light of this, in 1986 the Athena Trust organised a pioneering two-day conference in London titled 'The Bio-Revolution: Cornucopia or Pandora's Box?' The Trust invited an international group of genetic engineers, veterinarians, animal welfare campaigners, ecologists, regulators, philosophers, theologians, farmers and industrialists.[154] Then then raised questions about and debated the issues surrounding technology, science, agriculture and animal sentience from a range of standpoints, including the ethical, theological, philosophical, social, and environmental. It was the first

151 Joan Watson letter in *Agscene*, December 1987, p.8.

152 Joan Watson letter in *Agscene*, December 1987, p.8.

153 Peter Roberts, *Agscene*, May 1986, No.83, p.2.

154 *The Bio-Revolution: Cornucopia or Pandora's Box?* eds. Peter Wheale and Ruth McNally (Pluto Press, 1990), Preface, p.1.

ever international conference on genetic engineering and importantly it was intended for a non-scientific audience. Peter felt that previous conferences of its type had all been in-house affairs full of scientific terminology that was 'impossible for the ordinary person to understand'. A book of the conference was later published by London based Pluto Press and edited by Peter Wheale and Ruth McNally. Its readership was intended to be 'students of life-sciences and the humanities, and [it] should prove useful to everyone concerned about animal welfare and the environment'.[155]

At the conference, speakers debated subjects ranging from growth hormones and antibiotics, to the ethical consequences of genetic engineering, to issues around animal patenting and cloning. Speakers included Dr Michael Fox, vet, author and Vice President of the American Humane Society; Dr Andrew Linzey, psychologist and activist Richard Ryder, (creator of the term 'speciesism'); Andrew Lees, key campaigner with Friends of the Earth; Caroline Murphy, geneticist and RSPCA education officer; Richard Deakin, the marketing manager for agricultural-chemical giant Monsanto; and Dr Grahame Bullfield, an expert in transgenic animal research from the Edinburgh Roslin Institute.

In the mid-eighties Monsanto was still notorious for its creation of Agent Orange, used by the Americans during the Vietnam War to defoliate trees and shrubs and kill the food crops that were providing cover and food to the enemy. The spraying, however, did much more than kill off foliage, having been linked to the following disorders and health complaints in the Vietnamese people: chronic B-cell leukaemia, Hodgkin's lymphoma, multiple myeloma, non-Hodgkin's lymphoma, prostate cancer, respiratory cancer, lung cancer, Type 2 diabetes, immune system dysfunction, nerve disorders, muscular dysfunction, hormone disruption and heart disease.

> According to the Poison Papers, a public data trove that contains over 20,000 files about the chemical industry, although Monsanto and other manufacturers of Agent Orange knew about the dangers it posed to humans, they concealed this information before supplying it to the U.S. government.[156]

The herbicide Roundup, created by Monsanto and currently the world's most popular weedkiller, is chemically similar to Agent Orange. Also, Monsanto was one of the main manufacturers of DDT.

---

155 *The Bio-Revolution: Cornucopia or Pandora's Box?* eds. Peter Wheale and Ruth McNally (Pluto Press, 1990), Preface, p.1

156 https://asia.nikkei.com/Opinion/In-Vietnam-Monsanto-is-guilty-until-proven-innocent

The Roslin Institute states on its website that it 'aims to enhance the lives of animals and humans through world class research in animal biology'. It is worth pointing out that a few years after the Cornucopia conference it was this institute that infamously created Dolly, the world's first cloned sheep. Dolly was euthanised at six years old (less than half the natural life for a sheep) as she was riddled with joint and lung problems.

Although Peter was opposed to the kind of biotechnology that both Monsanto and the Roslin Institute supported and profited from, he was willing to engage both organisations in debate and to invite them to his conference. He referred to Roslin's research as 'mucking about with life itself', and with his years as a farmer and soil chemist behind him felt that companies like Monsanto were highly dangerous.

He was of course strongly opposed to the chemical-reliant monocultures that Monsanto forcefully sold. He questioned those present at the conference as to whether they felt it was 'right that future generations of farms and laboratory animals should be transformed into different and probably pathetic species which are protected by patents owned by the multinational companies'. Peter's debating style was spirited and emphatic, but as former Compassion CEO and friend Joyce D'Silva noted, 'Peter never vilified individuals, only institutions.'

1988 saw the creation of the so-called 'superpig' which *Agscene* featured on its front-cover; inside the magazine, the animal's official description read more like an advertisement than a description of a living creature: the pig is 'designed to be super-fast growing and with better than ever performance'. The side-effects of this 'design', however, as Peter noted, were 'acute arthritis, which meant the animal preferred to spend its life-sentence lying down'. Healthy pigs are active, inquisitive and social animals and are at least as intelligent as dogs. Other 'bio-disasters' of the eighties and nineties documented in *Agscene* were 'transgenic mice with deliberately deformed joints and sheep and chicken double their usual size.'

Cloning has always been, and continues to be, a contentious issue; it was of course a key feature of the Cornucopia debate. The purported aim of cloning animals, as the Roslin Institute's website writes today, is to create

> potentially life-saving medicines, ones that could only or easily be produced in the milk of cloned animals. The scientists at Roslin have managed to transfer human genes that produce proteins into sheep and cows, so that they can produce, for instance, the blood clotting agent factor IX to treat haemophilia

> or alpha-1-antitrypsin to treat cystic fibrosis and other lung conditions.

However, as inserting 'these genes into animals is a difficult and laborious process', cloning should mean that they only have to carry out this process once, as 'the resulting transgenic animal would build up a breeding stock[157] full of these 'life-saving medicinal agents'. Since Dolly's creation in 1996, years of medical research has resulted in the cloning of many more animals and a great deal of animal suffering, yet at the time of writing no medicine from any of these cloned animals has been 'permitted for use in humans'.[158] What Roslin's research *has* created, however, is 'a market for commercial services offering to clone pets or elite breeding livestock ... with a $100,000 price-tag'.[159] At the conference Peter lamented how humankind constantly assumes 'that if a line of research holds any hope of being in the immediate interests of mankind, no matter how frivolous or short-term, then it is acceptable' and that we somehow blind ourselves to questioning 'the welfare of the species involved'.[160]

The health hazards and ethics of the use of growth hormones on farm animals and their potentially perilous role in the food chain was the next topic up for debate. The synthetic production of bovine somatropin, BST, a peptide hormone made in a cow's pituitary gland, was a hot topic. BST was first synthesised artificially in the seventies, and was the first commercial product for agriculture from (you guessed it) Monsanto, the first company to get approval from the US Food and Drug Administration for its use in dairy cows. Peter and Compassion campaigners were deeply concerned at this potentially slippery slope of hormone and antibiotic use and abuse on farms worldwide.

The growth hormone BST is appealing to the factory farmer as it allows the dairy cow to produce much more milk than she ever naturally could; however, though BST is a naturally occurring hormone in cows the synthetic high doses of the BST chemical have countless side-effects; genetically engineered, injected BST pushes the dairy cow to her physical limit and increases problems such as painful mastitis and bacterial infections. Antibiotics are then required to clear up the infections, and the cycle continues, all the while pumping up the revenue stream of the chemical and pharmaceutical giants.

157 http://www.animalresearch.info/en/medical-advances/timeline/cloning-dolly-the-sheep/

158 Ibid.

159 Ibid.

160 *The Bio-Revolution: Cornucopia or Pandora's Box?* eds. Peter Wheale and Ruth McNally (Pluto Press, 1990), p.200.

Beyond the cow herself human overconsumption of BST, via dairy in the diet, has been linked to diabetes and some forms of cancer.[161] BST is currently still widely used on the American dairy farm, as well as, in more recent years, beef farms. Fortunately, the hormone has in recent years been banned throughout the EU and in some other countries, including Canada, Australia, Japan and Israel. The EU ban can in large part be credited to the tireless work of the London Food Commission's (LFC) BST Working Party, set up in 1987, which Compassion, amongst 15 other members, including the Women's Institute and the Maternity Alliance.

Yet despite Peter's disdain for both animal cloning and the profligate use of growth hormones on farms, he was keen to make clear that Compassion's stance was neither anti-science nor anti-progress. He applauded technological advancement that was humane and beneficial to people and planet, and he quoted Dr Leach of Kings College, Cambridge who in 1968 stated in the second-ever edition of *Agscene*, 'It is men who are blameworthy not science, science in itself is neutral. In the hands of men of goodwill it intensifies understanding and connectedness but in the hands of the sick it is an instrument of violence and alienation.'[162]

Peter believed in applying a balance of practical intellect and emotional intelligence, and in 1971 when Minister of Agriculture James Prior had suggested that it was necessary 'that the practice of science be separated from the business of ethics' Peter had responded with poetic concern, writing, 'Oh, what a dangerous situation this is, for the intellect, separated unnaturally from man's sense of justice, can destroy all beauty and all that our civilization has so painfully built up.'[163]

In another early newsletter, Peter took as an example of this separation a Cambridge University experiment in which sheep had had plastic bags inserted into their stomachs which were then inflated in order to ascertain whether the sheep's appetites could be reduced by a feeling of bloat; an 'infliction of suffering', Peter noted, which is a 'direct result of switching the emphasis in the teaching of agriculture to "pure science" '. He noted with dismay that 'agricultural students now spend two years studying "pure science" before they apply it to agriculture in their third and final year'.

This emphasis was a decided shift from Peter's own years of agricultural

161 World-Wire. What's in Your Milk?: https://www.organicconsumers.org/news/whats-your-milk-expose-dangers-genetically-engineered-milk and Geraci MJ, Cole M, Davis P. New onset diabetes associated with bovine growth hormone and testosterone abuse in a young body builder. Hum Exp Toxicol. 2011 Dec;30(12):2007-12. doi: 10.1177/0960327111408152. Epub 2011 May 9.

162 Dr Leach, *Agscene*, 1968 January/February No.2, p.4.

163 Peter Roberts, *Agscene*, August 1971, No.15, p.6.

study and practical work experience at Harper Adams agricultural college two decades earlier. Again, the influence of Ruth Harrison can be heard in Peter's arguments, who in her introduction to 1964's *Animal Machines* wrote how

> [t]he factory farmer cannot rely, as did his forebears, on generations of experience gained from the animals themselves and handed down from father to son; he relies instead on a vast array of backroom boys with computing machines working to discover the breeds, feeds and environment most suited to convert food into flesh at the greatest possible speed, and every batch of animals reaching market is a sequel to another experiment.[164]

Peter closed the Cornucopia/Pandora conference with a practical suggestion of developing a 'Blueprint for a Humane Agriculture'. In this speech he acknowledged that genetic engineering had the

> depressing quality of creating both utopian and dystopian extremists, and asked who is to say where the line is drawn between dystopian extremism and a proper concern? In the case of the regulation of genetic engineering, the usual political habit of compromise may mean the elimination of half a dozen species, perhaps our own included!

He mourned how

> [e]arly in this century we had an ecologically sound system of rotational farming that had, built into it, pest control, weed control and disease control … a system in which all wastes were returned to the soil for recycling, in which there was no chemical pollution nor organic pollution of natural resources. It was a system in which the animals and poultry were allowed to build up a natural relationship with each other in the herd and flock and in which they were given a tolerably square deal before they were killed [a system in which] certain taboos were built into farm tenancy agreements: no land should have the same

164 http://www.ru.org/index.php/animal-rights/113-treatment-of-animals-in-agriculture#references

> crop for two years running, except pasture; no hay or straw should be sold off the farm unless it was essential – in which case it had to be replaced by a succeeding crop or green manure; no straw must ever be burnt, and … no pregnant animal should be sent for slaughter.

He suggested that humankind's relationship to the natural world had unfortunately since the mid-forties been a progressively regressive one in which for 'perhaps forty thousand years, the human race has exploited animals for its own purposes, first as hunter, then as shepherd and more lately as tyrant'.[165]

165 *The Bio-Revolution: Cornucopia or Pandora's Box?* eds. Peter Wheale, Ruth McNally (Pluto Press, 1990), p.196.

15

SENTIENT BEINGS

> Whenever people say 'We mustn't be sentimental,' you can take it they are about to do something cruel. And if they add 'We must be realistic,' they mean they are going to make money out of it.
>
> Brigid Brophy

Other than the flourish of his trademark polka-dot bow tie and white pocket-handkerchief, Winston Churchill was dressed sombrely to address the crowds gathered outside the University of Zurich a year after the end of World War II. Both spiritually and materially, Europe was at a wretched low. But on the balcony of Switzerland's largest and most progressive university, Britain's lionised prime minister gave a characteristically rousing oration; speaking directly to the war-battered peoples of the Continent, he insisted on the post-war need to form a 'European Family'. His speech closed with his now famous bugle-like phrase 'Arise, Europe!'

The 1951 Treaty of Paris turned Churchill's utopian 'United States of Europe' into action, forging deals in which many former enemy countries now shared the key resources of coal and steel; resources previously central to the war effort. The treaty promised to ensure peace and prosperity and to reconstruct the floundering economies of the continent; its central tenets were focused on commerce and the development of the Common Market. Its offspring, the Treaty of Rome, created six years on in 1957, as well as focusing on steel and coal recognised the need for common transport and agriculture policies across the EU. The original member states – France, West Germany, Italy, Belgium, the Netherlands and Luxembourg – signed the Treaty in the appropriately grand 19th-century halls of the Palazzo Caffarelli-Clementino in Rome. A year before Britain officially joined the EU, in 1972, British Prime Minister Edward Heath travelled to Brussels and joined the other six members of Europe in signing the Treaty. Sixteen years on, in 1988, Peter made the bold statement that Compassion must get one specific clause of the Treaty of Rome altered; the clause that made no distinction between an inanimate, non-sentient trade good and a living animal.

At his desk at the Compassion HQ in Petersfield, Peter was in discussion with campaigns director Carol McKenna and education manager Joyce D'Silva. They were discussing Compassion's ongoing campaign against live animal exports. After over 20 years campaigning on this issue Peter, Carol and Joyce were frustrated by their slow progress. He took the Treaty booklet off his bookshelf as they spoke.

'This is *ridiculous*, animals being considered as goods,' said he.

'It's because of the Treaty that we can't get the British government to ban live exports,' said Carol and Joyce. They knew they needed a new status for animals in order to be able to argue that they shouldn't be treated in the same manner as light bulbs, tin cans, fruit and vegetables, or any other non-sentient good.

'In that case we've got to change it. Animals needed to be reclassified in the Treaty, their status changed from goods to sentient beings,' said Peter. Sentience is the capacity to feel, perceive, or experience subjectively. It is essentially defined as the ability to suffer.

In the late eighties Compassion were working on multiple campaigns, but their mission to get a full ban on live animal exports was their biggest fight of the era. Live export is the practice of transporting live animals, mainly sheep and cattle, across countries, even continents. The journeys are often long and the animals, unable to move, are deprived and stressed. At that time live export was big business, and one that traded from numerous ports in the UK. Trucks and ships were often overcrowded and poorly ventilated, and death in transit was common owing to crushing, overheating and dehydration, supplies of water and food on these long journeys often being totally inadequate.

Live exports vary in journey times, the shortest perhaps 8 hours and the longest, reported by Compassion in 2012, lasting over 60 hours. The animals leave the UK to end up in either the continent, Middle East or the antipodes, and on arrival at their destination are either fattened for later slaughter or immediately slaughtered. Both outcomes lack the protection of UK or European agricultural law, and often take place in countries where animal welfare laws are weaker than those of Europe.

Compassion had been working to get a total ban on live exports ever since the Roberts had founded the charity. A decade before Compassion's decision to challenge the Treaty of Rome Peter had warned what joining the EEC could mean for animal welfare. A 1972 edition of the *Lancashire Evening Post* reported the handing in of a Compassion-led petition which contained half a million signatures, more than 15,000 of them from the North West, which called on the 'government to ban the export of live animals to the continent before Britain's entry into the common market in three weeks' time'.

This time-sensitive petition was handed to a Mrs Peggy Fenner, parliamentary secretary at MAFF, and that was followed by a silent vigil outside the ministry and a protest outside the House of Commons. The *Lancashire Evening Post* reported that

> [o]rganisers of the protest Compassion in World Farming say that the Government could ban the export of live pigs, sheep and cattle to Europe overnight by cancelling meat export licences, and the organisation has documents to prove that exported animals are frequently maltreated en route to continental slaughterhouses and often cruelly slaughtered without being painlessly stunned when they arrive.

This story was also featured in the *Huddersfield Daily Examiner*, the *Wolverhampton Express and Star*, the *Southern Evening Echo* and the *Jersey Evening Post*.

The petition was urgent, and Peter made it clear that

> immediate action would be needed to stop the practice of live exports because once Britain is inside the EEC it will be obliged to abide by the Treaty of Rome which insists on free passage of livestock between member countries.[166]

Parliamentary secretary for agriculture Peggy Fenner, one paper reported 'refused to consider the ban but agreed to look into certain allegations'; another reported that she asserted that though the 'government would like to see more exports of carcass meat and fewer exports of live animals a total ban would not be practicable'.[167]

166 *Lancashire Evening Post*, Preston, 13.12.72.

167 *Express & Star*, 13.12.72 and *Huddersfield Daily Examiner*, 14.12.72.

Peter's prediction proved correct; once Britain had entered the EEC on 1 January 1973 it became even more difficult to encourage ministers to consider a live export ban, restrained as they were by other EU member states and the Treaty itself. Peter warned the public in 1974 that soon

> your MP will be able to vote whether to give live exports the green light or to insist on a total and permanent ban[168] ... the terrible part about it is that so few MPs will really know the facts and that many think vaguely that this subject, in which emotions run high, is very complicated and that they'd better rely on the recommendations of the O'Brien Report.

The 1974 O'Brien Report had recommended 'welfare safeguards and provisions', but not a total ban on the export of live animals, maintaining that there was 'no justification for a permanent ban on live exports on either humane or economic grounds'. The report had been filed after the committee, led by Lord O'Brien of Lothbury, former governor of the Bank of England, had been taken on what Peter described as 'a red-carpet tour' of the live export system, whereby the committee witnessed 'commercial animals being treated like high-value pedigree stock.' Despite huge public concern over the trade, most MPs in 1974 chose to follow the advice of O'Brien, and voted for live exports to continue. Comments from agriculture minister Fred Peart at a House of Commons sitting highlighted the division between welfarist groups, public feeling and supporters of the O' Brien Report, stating that

> It is clear that these findings of the O'Brien Committee point to a state of affairs far less disturbing than some would have us believe, and I hope they will be heeded by those who have sought to stir up public emotion on the more flimsy of the allegations.[169]

In contrast Peter, Compassion, and numerous other animal groups found O'Brien's report to present a highly misleading picture of the everyday reality of live exports – or, as Peter so wonderfully put it, a 'coat of whitewash, or a case of bringing out the best tea-service for the vicar'!

Yet live export was a contentious issue not only in terms of animal welfare but also in terms of economics, despite occasional accusations of impracticability, such as those by Minister Peggy Fenner.

168 Peter Roberts, *Agscene*, 1974, no page.
169 http://hansard.millbanksystems.com/commons/1975/jan/16/animals-for-slaughter-export

Interestingly, Peter was commonly described as a pragmatist by those on *both* sides of the table. For example, as an ethical vegetarian who actively promoted this diet, he was willing to appeal to farmers from an economic viewpoint when it came to live exports. In 1975 Compassion had run multiple ads in *Farmers Weekly* stating, 'BAN LIVE EXPORTS: SUPPLY THE HOME MARKET'. The ads continued:

> If you think that the export of livestock to foreign slaughterhouses should be banned, meat imports controlled, and the British farmer given first call on the £4 million per day British meat market, write to Compassion in World Farming.[170]

He appealed to British farmers to back the campaign on the grounds both that live exports were cruel and that the banning of them would give British farmers more of a square deal.[171] He appealed to farmers' feeling for their animals as well as their pockets, writing in *Farmers Weekly*:

> If ever there was a case on account of hardship for import control it is in the meat-producing industry. Each and every day this country imports £2 million worth of meat and meat preparations, staggeringly £750 million a year. That is the size of the home market which is denied to the British farmer. Yet the National Farmers Union is silent on the matter. Why? Does it have too much invested interest in meat-importing through its investment in FMC?* If not then why does the NFU not take action for the protection of its members? It is absurd that farmers should be reduced to craving the dry crust of live exports when the home market could offer them cream cakes.[172]
>
> (The FMC is the Food Machinery Corporation, an American chemical and amphibious tracked landing vehicles corporation.)

* The FMC is the Food Machinery Corporation, an American company that produces chemicals and amphibious tracked vehicles.

Compassion's political and legal director Peter Stevenson recalled a case he worked on with Peter which highlighted both the financial illogicality of live

170 *Farmers Weekly*, 5.12.75, No.11.

171 *East Anglican Daily Times*, 26.11.75.

172 'Go for the Cream and not the Crusts', letter, *Farmers Weekly*, 21.11.75.

exports and the absolute need for a change in the law, with the word 'sentience' to be included in it. In 1990, angry French farmers, suffering economically due to drought and declining meat prices, began attacking the lorries full of live sheep entering France from Britain. As Stevenson remembers, 'they felt it was undermining their livelihood and began attacking lorries, in the worst cases setting fire to them with the animals still trapped inside'. Stevenson recalls that it was 'widely documented in the media that the animals were being attacked', yet the Ministry continued to grant licences for these exports, even though the 'law said before granting a licence one must be sure animals were properly cared for'. Any law to protect farm animals seemed to be regularly contradicted and undermined by the overarching and statutory wording of the Treaty of Rome.

In 1988, after years of political machination as well as a change of government from Labour to Conservative, Peter asked how Compassion would ever be able to get live exports banned when legally there was considered to be 'no difference between a calf and a cabbage'. Legal director Peter Stevenson began looking more closely at the language of the Treaty. He noticed that it contained an annex 'in which a list of agricultural products included "live animals" alongside vegetables, cereals, meat and "guts, bladders and stomachs of animals" – as if the living creature were no different from the dead one'.[173] Stevenson issued a statement which said that 'just as the Community has taken on board a variety of issues of human rights and social justice, so now it is time for the Community to accord to our animal population a quality of life appropriate to our civilisation'.

Peter acknowledged that 'radical changes' (which was how the proposed ban on live exports was seen at the time) would not 'come quickly', noted that 'it took years to stop slave trading and the sending of little children up chimneys but … I think we'll win in the end'[174] and stated that until then live exports would remain a 'cruel, unnecessary and shameful … blot on Britain's reputation'.[175]

The general public agreed, and due to protests led by Compassion and other animal groups the trade received terrible and relentless press throughout the decade. There was a massive letter-writing campaign, millions of signatures on petitions across the country in support of the ban as well as the visual effect of weekly mass protests at ports, towns and airports across the UK. The daily press on live exports was so negative that in 1994 British/American company P&O Ferries, at that time the biggest ferry operator in the world, banned live

173 http://www.europarl.europa.eu/hearings/19951018/igc/doc63_en.htm
174 *Southern Evening Echo*, Southampton, 9.12.72.
175 Gerald Rafferty, 'Ireland's Saturday Night', 18.10.75

exports via its ships and added a detailed welfare policy to its website titled 'Serving with a Conscience'. The other big ferry companies of the era, including Stena Sealink and Brittany Ferries, followed suit.

Farmers wishing to continue with live exports had to charter their own ferries and planes, and started trading out of hitherto little used ports such as Shoreham-by-Sea and Brightlingsea, and small airports such as Bournemouth and Coventry. Exports now had to pass through small towns where streets were narrow and the animals were visible to residents – something which had not been the case with the big ports like Dover, through which exports could pass with near-invisibility; now you could see the animals' eyes through the slatted sides of the lorries as they passed.

In 1995 *The Guardian* reported that the 'protests at Brightlingsea continued daily, attracting up to 1,000 protestors each day'[176] and that one of the most striking aspects of these protests was the demographic of protestors. Writer Robert Garner noted that they were local people, 'many of whom had never protested about anything in their lives before'.[177] Other papers noted that the press and farming establishment were 'taken by surprise by the intensity of support and the presence of so many pensioners and mothers with young children, which challenged the stereotype of the typical animal rights protester'.[178] Instead of the usual suspects of 'Earth First campaigners, Greenpeace and animal rights groups, they found middle class, adult protesters alongside and outnumbering the radical young'.[179]

In February 1995, a year after P&O's ban, a small group of protesters demonstrated at Coventry airport, where calves were being flown to Amsterdam for distribution across Europe. Importantly, at that time veal crates were still legal in parts of Europe. There were just 35 protesters that day, 10 of whom broke through police lines to try and halt the export lorry. Among them was 31-year-old Jill Phipps, who in the course of the protest was knocked down by a lorry and crushed beneath its wheels.

Phipps, inspired by her parents, had been an animal activist since childhood, protesting the fur trade aged eleven and soon after that going vegetarian, then convincing the rest of her family to follow suit. As a teenager she'd been a member of the Eastern Animal Liberation League and as an adult she attended foxhunt sabotage demos and meetings. She protested against

176 Michael *Durham 'Animal rights: Out of the kitchen, into the firing line' The Guardian, London, 17 September 1995, p.12.*

177 Robert *Garner, Animals, Politics and Morality.* (Manchester University Press, *Manchester, 2004) p.72.*

178 Benjamin *Seel, Matthew Paterson, Brian Doherty, Direct Action in British Environmentalism (Routledge, London, 2000).*

179 Tessa *Muncey, Creating Autoethnographies (Sage Publications, London, 2010). p.207.*

Dutch-British food and cosmetic giant Unilever's use of animal testing, and received a suspended sentence for damaging property at Unilever's soap factory at Port Sunlight in Merseyside. The sentence had been suspended because she was pregnant at the time.

The month before the Coventry airport protest, Phipps had spent her 31st birthday demonstrating outside the home of Christopher Barrett-Jolley who ran Phoenix Aviation, the company that exported the calves from Coventry airport.[180]

Phipps suffered such severe spinal injuries from the lorry that she died on the way to hospital. Her death received huge publicity and was raised that week in the House of Commons, in John Major's PMQs. Leading livestock exporter Richard Otley claimed Phipps was 'largely to blame for her own death', while fellow protesters claimed her 'a martyr to animal liberation'.[181] Her close friend and fellow animal rights campaigner John Curtin noted how 'if you get hurt in the line of demonstrating you're seen as some radical ... So there were these two characters – the respectable businessman Christopher Barrett-Jolley and the violent protester – but you have to flip that on its head to see the reality.'[182]

In 2002 Barrett-Jolley was jailed for cocaine-smuggling to developing countries; it was alleged, too, that he had been gun-running.[183]

But back in 1995, it was for the first time ever that animal welfare made headline news every day, and for months on end. In consequence live export numbers dropped significantly: in the mid-nineties over 2.5 million animals were transported live from the UK to be fattened and slaughtered overseas; by 2010 the number had dropped to 4,000 sheep and no calves.[184]

Since 2010, however, the number of live animals exported from the UK, mainly through Ramsgate and Dover, has been on the increase despite the widespread opposition of the local community, the British public and a number of MPs. In 2011 and 2012, over 130,000 live animals were transported from ports in the south-east of England.[185] In addition, there is a substantial export trade in young calves from Northern Ireland to the continent.[186]

Beyond live exports, the wording in the Treaty of Rome was a blockade to every campaign that Compassion, and other European animal groups, large

180 http://www.independent.co.uk/news/uk/for-what-cause-did-jill-phipps-die-1571300.html

181 http://www.independent.co.uk/news/uk/for-what-cause-did-jill-phipps-die-1571300.html

182 ibid.

183 http://www.coventrytelegraph.net/news/coventry-news/jill-phipps-remembered-20-years-8551338

184 https://www.ciwf.org.uk/our-campaigns/live-animal-transport/uk-live-exports/

185 https://www.conservativeanimalwelfarefoundation.org/latest-news-updates/our-article-on-conhome-29th-august-2016-animals-are-not-freight/

186 https://www.ciwf.org.uk/our-campaigns/ban-live-exports/

and small, were working on so tirelessly. If , according to European law, farm animals were not capable of suffering, then EU farmers, animal transporters and abattoir workers were legally free to treat the animals in their care as they would any object. Carol McKenna visited the other major animal welfare organisations of the day and asked for their backing in the campaign to get the Treaty of Rome altered in order to recognise animal sentience. She approached the RSPCA, Eurogroup for Animals and WSPA (**World Society for the Protection of Animals**).

The various charities' unanimous response was that Compassion ought to be more 'politically realistic' and the attitude was one of general condescension. Even Ruth Harrison, whose seminal and galvanising book *Animal Machines* had inspired Anna and Peter, suggested that attempting to get the Treaty changed was 'a waste of time'. There were echoes of history repeating itself two decades on from Compassion's founding; back then, Anna and Peter had approached the prominent animal charities of the day and asked that they add farm animals, rather than only companion and wild animals, to their agendas, and the response had been similar – one of either silence or accusations of impracticability. Now this was clearly happening again in relation to live exports.

Undeterred, Peter, Joyce, Carol and Philip Lymbery took their petition to campaign groups across Europe. Thankfully many of the European groups were more supportive, translating the petitions into their country's language, which they then took to their MEPs, the majority of whom were pleased to back the bill.

In just over three years, between 1988 and 1991, Compassion obtained one million signatures requesting the Treaty be changed. Peter Stevenson, on one of his first days in the Compassion office, and Philip Lymbery drove across the continent to deliver the boxes of petitions to the European Parliament in Strasbourg. At that time, it was the largest petition the EU had ever received. NB although the internet had been invented in the early eighties it was not generally available until around 1994. Hence the huge task of gathering that number of signatures on paper.

Throughout the nineties battles continued to rage over both the ethics and economics of the live export trade, added to which , due to the BSE crisis, was the safety of British beef, and Compassion continued to campaign hard to get the Treaty of Rome altered. This seemingly small focus on the altering of a few phrases within the Treaty, as Peter and Compassion knew, could prove monumental in terms of real world improvement for the lives of millions of animals. As Rudyard Kipling put it: 'Words are, of course, the most powerful drug used by mankind.'

Despite the masses of signatures handed in at Strasbourg in 1991 requesting that animals be reclassified as sentient beings it took another five years of active campaigning before legislative change came in the UK. In 1994 the petition was endorsed by the European Parliament, and in 1996 high-profile Compassion patron, actor and national treasure Joanna Lumley delivered yet another petition supporting the ban, this time to Conservative Prime Minister John Major. The petition asked that the government 'secure the status of animals as sentient beings at the forthcoming Intergovernmental Conference on Reform of the Treaty'.[187]

That same year Compassion led a large demonstration outside the EU Prime Ministers' conference in Amsterdam, calling for greater recognition of animal sentience in European Law. The years of campaigning finally paid off, as the EU Prime Ministers added 'the protocol recognising animals as sentient beings to the European Treaty',[188] a change which **'brought a greater opportunity to improve the lives of billions of individual animals, each one a sentient being with intrinsic value'.[189] The re-classification of animals as sentient beings within European law is viewed by many as Compassion's greatest achievement to date, and in 2006 *The Times* hailed it as 'a remarkable achievement that is now bringing about a profound change in the way that animals of all kinds are regarded'.**[190]

187 Peter Roberts, 'Go for the Cream and not the Crusts' letter, *Farmers Weekly*, 21.11.75.
188 http://www.ciwf.org.uk/news/2009/12/the-lisbon-treaty-recognising-animal-sentience
189 http://www.ciwf.org.uk/news/2009/12/the-lisbon-treaty-recognising-animal-sentience
190 *The Times*, 27.11.06, p.50.

16

A BERNARD MATTHEWS-FREE CHRISTMAS

In *Eating Animals*, Jonathan Safran Foer, novelist and vegetarian activist, examines the multiple strongholds of tradition, family and convention surrounding food, strongholds that –in the West at least – come to the fore on the hallowed occasions of Thanksgiving and Christmas Day:

> Thinking about eating animals, especially publically [*sic*], releases unexpected forces into the word. The questions are charged like few others. From one angle of vision, meat is just another thing we consume, and matters in the same way as the degree of consumption of napkins or SUVs … Try changing napkins at a Thanksgiving though … and you'll have a hard time getting anyone worked up. Raise the question of a vegetarian Thanksgiving and you'll have no problem eliciting strong opinions - at least strong opinions. The question of eating animals hits chords that resonate deeply with our sense of self - our memories, desires and values. Those resonances are potentially controversial, potentially threatening, potentially inspiring but always filled with meaning.[191]

In eighties Britain a catchphrase thought up by a highly-successful multimillionaire turkey farmer became common parlance. 'Bootiful,' said Bernard Matthews in his thick Norfolk accent on his prime-time TV commercial advertising turkey breast roast, and the phrase entered the nation's consciousness. Founded in 1950, Bernard Matthews' farm became by far the

191 Jonathan Safran Foer, *Eating Animals* (Penguin Group, London) 2009. p. 264.

largest turkey farm in the country, with outposts in Germany and Hungary. The vast majority of the millions of birds on his farms were factory farmed and Matthews was awarded a Queen's Service Medal by the Government of New Zealand in 1989.

It was Matthews who in 2005 invented the turkey twizzler, which came to be seen as a symbol of much that was wrong with the modern, processed-meat-heavy diets, in large part due to Jamie Oliver featuring the food on his TV show 'Jamie's School Dinners'.

In 2006 the company hit the headlines again when two contract workers were convicted of animal cruelty after members of Hillside Animal Sanctuary had covertly filmed them playing 'baseball' with live turkeys. Matthews was due a CBE in 2007, but this was postponed due to an outbreak of avian flu at one of his farms. On his death in 2010, the company was sold for £87.5 million.[192] Today around 10 million turkeys are consumed in the United Kingdom over Christmas, the vast majority of them intensively reared.

An intensively reared turkey's life is a similarly miserable one to that of the broiler chicken, and the turkeys share the broiler chickens' ill-health, including their painful bodily ulcers and burns caused by the overcrowding and often unsanitary living conditions. Like the broiler-chicken, the factory-farmed turkey is prone to broken bones and heart conditions owing to their unnaturally fast, genetically engineered growth rates. Additionally, turkeys have the caveat that given their inherent vulnerability to disease they are perhaps the worst fit of any animal for the factory farm model, so are given more and higher doses of antibiotics than any other farmed animal.

Compassion's consistent campaign message was that Christmas should be a time of goodwill to all living creatures – not just humans – and each year they asked the public and their supporters to boycott the traditional Christmas dinner for a meatless alternative.[193] Anna's three cookbooks offered several cruelty-free options for the festive table. They hoped more people would decide to celebrate Christmas without cruelty, and the front cover of the December issue of *Agscene* always featured a turkey tale. One year it showed a conveyor belt of turkeys hanging on a slaughterhouse line under the striking 'Christmas is About Peace Not Crucifixion'. Another cover had a couple putting a nut loaf and roast vegetables into an oven while a pet turkey roamed the kitchen, under the headline: 'Sanctuary or Slaughter this Christmas?' with an explanation that 'Clyde' was 'one of the lucky turkeys given a safe home last Christmas by the American group Farm Sanctuary'.

---

192 https://www.foodmanufacture.co.uk/Article/2016/09/21/Turkey-giant-Bernard-Matthews-purchased-by-Ranjit-Boparan

193

Besides its supporters' magazine, Compassion reached a wider audience in December 1986 when they were featured on BBC 9 o'clock news showcasing 'an alternative Christmas dinner'. For the feast, Compassion employees Joyce D'Silva and Carol McKenna, got cooking, while a long-term supporter, Hilary Nimmo, provided the setting and the guests. They dined well, starting with stuffed avocados, then a superb cashew and mushroom nut roast with sherry sauce and chestnut purée en croute, and ended up with Christmas pudding.

Peter's elder brother Frank, his wife Gill and their sons Barry and Mike had for years spent Christmas at Copse House. But one year, without any explanation, these visits stopped. Barry later said that he suspects this was because his father had had enough of Christmas without turkey. On the journey from Frank's family home in Birmingham to the Roberts' home in Greatham, he would warn his children 'not to ask for meat while they were down there – that was a big no, no,' that Anna would be 'cooking vegetarian food, and to avoid the subject of meat altogether.' When, at the end of a visit Frank and the family would be given boxes of Direct Foods produce, he would thank Peter and Anna very much for their hospitality, then assure Peter that he'd be visiting the first steak house he saw on the journey home. When Peter had become vegetarian and quit farming, the only job that he then knew how to do, and founded a charity for animals, Frank had thought his younger brother had 'lost the plot.'

As Frank's family visits ceased and the three Roberts daughters grew up and had their own children, the Copse House Christmas expanded again. By the nineties we were a close-knit family of 14 with aunties, uncles, cousins, partners jostling for space in the Roberts cottage. All but one of the Copse House regulars were vegetarian, and meat was never served inside the house even if a guest was omnivorous. In many ways Anna and Peter were traditional, and other than Compassion, family was the most important thing in both their lives so Christmas was a big deal.

We would arrive on Christmas Eve and stay until the 27th, and Peter was uncharacteristically selfish when it came to the season: in the few years that his daughters went to their husbands' families instead he would complain. We would arrive to find him midway up the 40-foot silver birch at the entrance to Copse House, wrapping fairy lights around trunk and branches; although he was almost 70 he was still as adroit a tree climber as he'd been in his youth.

At the kitchen stove one of the three Roberts daughters would pour red wine, orange juice and brandy into a large saucepan to mull with slices of orange, plus cloves and star anise, and the children would offload parcels and pile them under the tree, pine needles forming a halo around the base. There were always tree chocolates for us and decorations knitted by Anna, Peter had prepared well for cosyness, and a huge pile of seasoned logs overflowed the basket by the fire. In every window stood electronic triangle candelabra, complete with plastic wax drippings, then there were strings of cards, gaudy tinfoil decorations and tinsel hanging from every surface. We children were allowed one pacifying present on Christmas Eve, and then had to be in bed by midnight, making sure to leave a mince pie and a glass of sherry for Santa, plus a carrot for the reindeer. The back room, known as the children's room, was rammed with bunk beds and zedbeds, a chest full of toys and costumes, shelves of stories, including a huge Paddington tome, Cicely Mary Barker's beautiful Flower Fairy books, and lots of Dahl, and tucked in a corner was a dark wood wardrobe where the monster lived, along with old wedding dresses and a meditation pyramid.

The last decade of the 20th century was the era of personal camcorders, so some cute and cringeworthy scenes of Christmas were captured – including an abbreviated collage of the typical Roberts scene: Anna bursting through the bathroom door in a towel, singing 'If I Ruled the World'. Bristol Cream sherry shots. Gift-giving in the lounge: Anna and Peter's interior design is an eclectic mix of kitsch, country-cottage and boho, and every surface of the room vibrates with colour; the walls are racing green against red tapestry curtains, a large painting of horses running towards the pearly gates of heaven hangs above the fireplace, on another wall are gold and turquoise Egyptian prints, in the old bread-oven stand a few hunks of crystal and a bronze Buddha.

Anna and Peter give creatively: pottery wheels, science sets with Bunsen burners, basket-weaving kits and watercolours. A cacophony ensues, keyboards and a karaoke set as gifts. Anna and Peter's Alsatian roams through the paper, tinsel and ribbon. Wine is mulled. By early evening we're at the kitchen table with leftovers, dousing the pudding in brandy, match-ready. One year there's an elf dance party and a cocktail-making course followed by a Pilates plank competition, another year sumo wrestling in costume.

However, new friends, boyfriends, strangers at parties, only ever really wanted to know one thing about Christmas in a vegetarian household. *What did we eat? What was it like without turkey?* The implication, as Jonathan Safran Foer points out, is that 'the tradition is broken, or injured by a lack of meat'.[194] This specifically sanctioned meal at the centre of the day had, it seemed, come to epitomise the celebration as a whole.

At Copse House the Christmas cooking began, as in many homes, at 11 am; everyone began food prepping, an Elvis Presley Christmas CD was put on, festive paper and crackers were laid on the table, candles lit. About four hours later Christmas lunch was ready. The cooks were toasted, as were absent

194 Jonathan Safran Foer, *Eating Animals* (Penguin Group, London, 2009) p.251.

friends, and instead of a bird as centrepiece there would be four varieties of nut roast, plus the usual trimmings: potatoes (*sans* goose fat), boiled-to-death sprouts, carrots, peas, roasted chestnuts, sage and onion stuffing balls, bread pudding made using soya milk, gravy and Jumbo Grills, a product familiar to vegetarians of the seventies onwards– strange, dark brown lumps of TVP, which tasted much better than they looked. There were seconds and thirds. Christmas in a vegetarian household was unremarkable and predictable in the possible best way, full of the same ridiculousness, over-indulgence and ritualistic merriment as almost any other British home celebrating Christmas in the modern era.

It is worth noting, though, that while until 1961 Peter, by all accounts, used to love a Sunday joint, and that both Anna and Peter had grown up with traditional turkey or goose Christmas dinners in their family homes, the Robert's daughters and grandchildren had nothing to compare with or to miss, having never experienced a Christmas turkey dinner. However, neither Anna nor Peter ever voiced a wish for the more traditional Christmas meals they'd grown up with; for them it seemed that the annual Christmas celebration was, to use Safran-Foer's phrase, 'enhanced' rather than diminished by the lack of turkey at the family table. The choice not to eat the bird was 'a more active way of celebrating' the season.[195]

195 Jonathan Safran Foer, *Eating Animals* (Penguin Group, London, 2009) p.251.

17

THE NINETIES AND THE NOUGHTIES

Throughout the 20 years around the turn of the millennium, alongside their fervent campaigning around the wording of the Treaty of Rome, Compassion continued to push on key issues surrounding battery chickens and intensively farmed pigs, campaigns which Anna and Peter had begun in the Copse House Sunroom back in 1967.

Almost 25 years on, their efforts were tangibly paying off. In 1991 the European Union finally agreed to ban battery cages for egg-laying hens. Due to the sheer number of animals – in the UK alone we consume over 30 million eggs per day – that this ban has benefited, it has been hailed by many as the biggest victory for animal welfare in recent history.

More victories followed; in 1999 a bill to outlaw the keeping of pregnant sows in narrow stalls or on chains during their pregnancies was introduced in the UK, and in 2013 sow crates were banned across the EU. Constant petition campaigns and public demonstrations with Hetty, a human-sized hen (created for Compassion by the puppet-maker of *Spitting Image), helped achieve this ban.*

They also made a controversial cinema short starring David Graham (who had been the voice of the Daleks). In the one-minute screening, the conventionally suited Graham explains in the rational, calm, yet highly sinister tone and style of a government public service announcement in the style of Orwell's *1984*, that the audience should: 'not obstruct the gangways while the cages are installed' and that 'there is no cause for alarm' as 'these cages are for your protection'; furthermore that audience members should 'cooperate with the surgeons' who will be 'removing your teeth and nails' in order to 'reduce the incidence of cannibalism' – analogous to the debeaking process undergone by intensively reared birds. The close-up of Graham's face continues as he explains that though there may be some 'discomfort … your space allowance complies exactly with government regulations,' and that audience members should have 'the satisfaction of knowing that you are part of one of the world's most cost-effective production systems'. The film then switches to live filming inside a battery farm while Graham's reassuring voiceover continues, stating that 'this system has been tested on 45 million specimens – with, I might add, your approval'. The footage concludes with an image of a caged hen and the directive to Buy Free Range Eggs.

The screening was swiftly banned in cinemas, the authorities arguing that it would frighten cinemagoers too much; they might really believe they were about to be caged and trapped in their seats. (A ludicrous suggestion – but influenced, perhaps, by the fallout from Orson Welles' 1938 radio rendition of H.G. Wells' *The War of the Worlds*, which had generated panic among listeners.) The film was given an 18 certificate, and under pressure from the farming industry was banned by some commercial cinemas; however this prohibition fitted the maxim of 'no publicity is bad publicity' as more people got to see it; it made the breakfast news, when many were eating their eggs … The film, *Welcome to the Battery*, is still available on YouTube and is still screened in France.

The sow crate campaign was the other major success story of that era, and the simple yet effective campaign strategy included a massive, expertly orchestrated, letter-writing campaign to Members of Parliament. It was this which largely led to the ban, one MP revealing that 'he received more post on the issue than about the entire Gulf War'.[196] Once more, it's worth highlighting here that the internet and internet petitions were barely available to campaigners.

Alongside these victories and the bad press surrounding the live export trade, factory farming suffered yet another PR blow in the early nineties, when the BSE epidemic made the headlines – headlines which eventually directly linked the sub-standard practices of factory farmers to the outbreak of the disease.

BSE, bovine spongiform encephalopathy, is a disease of cattle affecting the animal's central nervous system and brain, causing it to behave abnormally. Infected cows usually appear agitated, stagger, and display violent behaviour. BSE is usually fatal; there is no known treatment for it and no known vaccine to prevent it. It's believed to be caused by an agent such as a prion or a virino (an infectious protein particle), and to be related to Creutzfeldt–Jakob disease (CJD) in humans.

---

196 *Daily Telegraph*, 4.12.2006. p.25.

Although the disease peaked in the nineties, the story of its development dates back to the 18th century; it was first noted as scrapie, a brain disease in sheep, in 1732. From there it developed into a fatal disease which could pass from cattle to humans. Though the twenties saw the first recorded incidence of BSE-related CJD, it wasn't until 1986 that the disease was officially recognised, and it wasn't until a year later that UK government ministers were told about this new disease. Meat and bonemeal were then identified as the 'only viable hypothesis for the cause of BSE'.[197] Although at first that might seem surprising, it had by that point become normal practice amongst factory farmers to feed slaughterhouse remains to their cattle, thus saving money. The period of intensification during the seventies and eighties is thought 'to be the period when scrapie "jumped" the species barrier and reappeared in cattle as BSE after changes in the rendering process'.[198]

But as journalist Monika Nickelsburg in *The Week* explained in 2013:

> contracting bugs from animals is nothing new, in fact, zoonoses (infectious diseases transferred between species) are a natural part of evolutionary biology … modern industrial farming practices can turn health issues that were once fairly benign into real concerns. Factory farming creates perfect conditions for the proliferation of super bugs. The stress and unsanitary conditions of CAFOs (concentrated animal feeding operations) weaken animals' immune systems, making them more susceptible to infection; overcrowding allows disease to spread quickly and easily; and over time, antibiotics can cause resistant strains of bacteria to evolve. These conditions, combined with a lack of diversification, create a petri dish for dangerous diseases.[199]

Mad cow disease, as it came to be commonly known, realised itself fully as 'a result of offal, a mixture of the organs and entrails of butchered cattle' being put into feed, and although farmers

learned that cannibalism can cause infectious neurodegenerative diseases in livestock, government ministers were at best slow to act, and at worst flagrantly ignored the warnings of farmers, concerned consumer groups and

197 https://www.theguardian.com/uk/2000/oct/26/bse3 BSE Crisis Timeline Thursday 26 October 2000, accessed 20.7.17.

198 https://www.theguardian.com/uk/2000/oct/26/bse3 BSE Crisis Timeline Thursday 26 October 2000, accessed 20.7.17.

199 http://theweek.com/articles/457135/5-modern-diseases-grown-by-factory-farming *The Week*, Monica Nickelsburg, 7.11.13 accessed 20.7.17

animal campaigners. The system of the factory farm industry had contributed to the rise of the BSE crisis yet agriculture ministers were reluctant to admit blame. So reluctant in fact that in 1990, then conservative agriculture minister John Gummer made great show of feeding his four-year old daughter Cordelia a British beefburger for a press shoot at a boat show in Suffolk. This spectacle was at the height of the scare and two years on BSE reached its peak with 100,000 confirmed cases.

In 1995 the 'first known victim of variant CJD, 19-year-old Stephen Churchill, died, and three more people died' later that year.[200]

Philip Lymbery, the current CEO of Compassion, who was working as the campaigns manager at the height of the BSE crisis, recalls daily morning phone calls from the press asking for quotes from the charity on the escalating epidemic. Finally, in March 1996, the government health secretary, Stephen Dorrell, officially announced that there was a 'probable link' between cattle disease and poor farming practices, and a month later the EC imposed a world-wide ban on all British beef exports. Some British schools banned beef, and beef-on-the-bone was officially banned by the government later that year.[201]

In October 2000 the British government released the results of the BSE inquiry. Peter had for years and years been predicting links between the sub-par factory farming practices and potentially fatal diseases in humans – and in more cynical mood he had prophesied that it would be 'disease rather than compassion that would eventually persuade the whole world to go vegetarian'.

200 https://www.theguardian.com/uk/2000/oct/26/bse3 BSE Crisis Timeline Thursday 26 October 2000, accessed 20.7.17.

201 https://www.theguardian.com/uk/2000/oct/26/bse3

18

SACRED SEWING

> All beings tremble before violence. All love life. See yourself in others. Then whom can you hurt? What harm can you do?
>
> The Dhammapada 129–130

The sewing room at Copse House, was where, amongst a mix of more ordinary objects, the more unusual and esoteric items dwelled; to me as a child this room held a strong fascination; as an adult it somehow seemed to be the material manifestation of Anna and Peter's eclectic spiritual journey.

It contained a mix of arts and crafts, exercise equipment, the sewing paraphernalia which gave it its name, and of course some spiritual curiosities. Then against one wall was the peach-coloured Solotone, a lazy person's exercise machine – you would lie down supine and press different buttons to lift and lower your limbs or torso – and there were shelves packed with colouring books, paints, felt pens, puzzles and pots of glorious-smelling playdough (that madeleine to instant childhood); there was a magical (to us grandchildren) lunchbox, full to the brim with coloured buttons that occasionally we'd be allowed to look through and select our favourites to keep; and the room was dominated by a top-of-the-range sewing machine and various sewing, knitting and crocheting paraphernalia. Anna had taught us all how to sew and knit with greater or lesser success and it was in this room that she and her close friend Betty Rabjohn knitted jumper after jumper; we received some of these chunky knits in rainbow colours with wooden toggle buttons, but most were sent to Children on the Edge, a charity for eastern European orphans started by Body Shop founder Anita Roddick in response to crises in that area in the early nineties.

Yet beyond playroom and humanitarian knitting club, the sewing room had a sacred quality. Amongst the arts and crafts was a wooden octagonal table that held a Tibetan singing bowl, ashy traces of incense and a hunk of tourmaline – the room held dozens of crystals in fact, as well as some petrified wood and fossils, and the bookshelves were lined with countless copies *Unconditional Love* (Anna would hand them out to anyone willing to take them), photographs of Sathya Sai Baba, and stacks of White Eagle Lodge magazines. Littered about the room there was also a copper mediation pyramid, some *baoding*, Chinese meditation balls, crystal light-refracting prisms (which cast rainbows indoors when held up into the light). Pendulums dangled from the window, and wooden boxes overflowed with more incense than a person could burn in an entire lifetime – cones, sticks, bricks, charcoal burners and Frankincense resin – and on the whitewashed wall hung a framed Serenity prayer (written by American theologian Reinhold Niebuhr and famed for its links to Alcoholics Anonymous). Another frame contained the opening lines of poet and radical Christian William Blake's 'Auguries of Innocence' (those lines about eternity and grains of sand), and yet another a photograph of a single set of footprints on a beach with writing overlaying suggesting that there was only one set of prints during a time of struggle because this was when Jesus carried the suffering mortal.

Strong religious and spiritual devotion has not been uncommon within the animal movement, and this devotion has often been one that has wavered from the UK's more mainstream doctrines of the Church of England and Catholicism. Lady Muriel Dowding, Founder of Beauty Without Cruelty, was, as mentioned earlier, a Theosophist. Friend of the Roberts and animal advocate Esmé Wynne-Tyson was a convert to Christian Science and believed in mediumship.

Esmé was an actor in the West End and a friend of Noel Coward; allegedly, aspects of the character Madame Arcati in his play *Blithe Spirit* were based on her. Peter enjoyed many of Esmé's books, including *The Philosophy of Compassion* and *The Return of the Goddess*. Esmé's son, Jon Wynne-Tyson, also a close friend of the Roberts and supporter of Compassion's work, followed in his mother's vegetarian, humanitarian and spiritual footsteps, publishing *Food for a Future: the Complete Case for Vegetarianism* (1979), and *The Extended Circle: A Dictionary of Humane Thought* (1984).

Jean Le Fevre, Anna's best friend – she who ran the wolf sanctuary in Texas – was, as mentioned earlier, a White Eagle Lodge minister, and Ruth Harrison, author of *Animal Machines*, was a Quaker.

Spirituality was a subtle but continual backdrop to time spent with Anna and Peter and to life at Copse House in general. I recall, around the age of eight, being sat at the kitchen table reading a book when Peter came in and told Anna a story which has stuck with me all these years.. He'd been driving into Petersfield and on the way up Church Lane had offered a walker a lift. The walker had accepted, and once in Peter's car had very quickly struck up a conversation about religion. The man was a resident at the L'Abri Manor, an evangelical Christian fellowship, located at the top of Church Lane. Peter's passenger asked him what he believed in, and Peter answered that he liked to 'take the best bits from all of them', which did not go down well with the man. Anna's response to her husband's answer was, 'Well, quite right.' Peter's ideology had more depth and complexity than this somewhat flippant answer implies, but in broad strokes this was his approach to the metaphysical. The spiritual aspect of life was incredibly important to him, but he was committed to learning across the breadth of belief systems rather than being committed to a single organised and institutionalised faith. His beliefs seemed profound but were never dogmatic or rigid.

As is so well documented as to have become cliché, the sixties was an era of intense spiritual self-discovery, so one might assume that the Roberts were simply following the trend of that decade and that this was when their interest in the non-material realms began. At that time religion began to be decentralised from the official Judaeo-Christian authorities dominant for so long in the West. Other ideologies, ranging from Taoism to neo-Pagan traditions, to Hinduism and Gnosticism, began to gain force in Britain. This was the era of vision quests and the aim was self-actualisation. The Beatles were very publicly hanging out with gurus in Rishikesh, India; sandalwood beads had become a fashion statement, and Maslow's hierarchy of needs, created in the forties, which focused on 'people's capacity for goodness and transcendence' was gaining massive popularity, as were ideas surrounding karma, reincarnation and ecological responsibility, linked to a spiritual connection to Gaia or Mother Earth.

Yet Peter's spiritual inquiry had in fact begun two decades earlier, in the early forties; in his late teens his bookstack had indicated an esoteric bent in him, and in 1943, at agricultural college, he had been a regular practitioner of yoga and meditation. Even so, during the mid-to-late sixties Peter's spiritual life did indeed bloom, as did Anna's.

During that decade and beyond, Peter immersed himself further in the study of world and mystery religions, while Anna had a leaning towards crystal healing, angels and gurus. The whole family had their horoscopes read by friend, popular vegetarian cookbook author and astrologist Rose Elliot.

They read books by Indian guru Shri Chinmoy and set up a weekly meditation circle based on his teachings, which they held in Peter's garden shed, a handful of local, like-minded friends joining them. Chinmoy appealed to the Roberts both as an ethical vegetarian and as a great proponent of meditation; he was one of the first individuals to establish meditation centres in the West – in New York City, in 1964 – and, interestingly, advocated athleticism as an aspect of his spiritual teachings, including distance running, swimming and weightlifting. This suited Peter well, as he liked to balance the physical and spiritual by swimming meditative lengths in the pool. (Peter had spent his inheritance from his father on a swimming pool, characteristically choosing to spend the money on something concrete and extravagant, rather than allowing the money to be absorbed by sensible domestic needs or to go into rainy-day savings.)

Peter's spirituality inevitably led his work for Compassion and in *Agscene* in the late sixties he wrote an article on the importance, in our increasingly dissociated world, of seeking 'self-orientation by devoting an hour or two a week to the study of the world's sacred writings'.[202]

It was this search through the 'sacred writings' for 'the best bits' that took Anna and Peter on some profound journeys, both physical and spiritual. To use writer and journalist Mick Brown's phrase, the Roberts were 'spiritual tourists', yet two key spiritual movements were to influence them more profoundly than many of the others they explored.

These were the teachings of the White Eagle Lodge and of Indian guru Sathya Sai Baba. As mentioned earlier, The Lodge had been an early catalyst for the Roberts' founding of Compassion, and was important in both their lives for over three decades. Though Peter had initially introduced Anna to The Lodge, she was to become more actively involved in it than him, even choosing to change her birth name by deed-poll from Edna to Anna in the early sixties according to advice from The Lodge. Anna is a name derived from the concept of grace, and in apocryphal readings from both the Bible and the Qur'an it was said to be the name of Jesus's grandmother, the Virgin Mary's mother. The name Edna, on the other hand, is associated with fire and pleasure!

But beyond The Lodge, Sai Baba was the most profound spiritual influence on their lives. In Peter's case the study of Hinduism, of the Bible, of White Eagle's teachings, of spiritual texts by Sri Chinmoy, Sai Baba and others, took on an academic bent; he mined and examined the many sacred texts of the world, looking for ethical guidelines and answers to life's unfathomable questions. Anna's interest, however, took on a much more personal and emotional tone: she sought a direct relationship with a higher force and settled on Sai Baba. His was a presence in Copse House for many years, and in casual conversation Anna would sometimes say 'Thank Baba' rather than 'Thank God'. As a child I

202 Peter Roberts, Agscene, January/February, 1968, issue 2.

had a framed picture of him in my bedroom that Anna/Nan must have gifted me. ('He's like the Indian Jesus,' I'd explained to my next door neighbour during a playdate.)

Born as Narayana Raju in 1926 into a regular Hindu family in the village of Puttaparthi at aged fourteen, Narayana claimed to be a reincarnation of Shirdi Sai Baba, an Indian spiritual leader who had died eight years previously. Both Shirdi and Sathya were interfaith teachers, though Sai Baba subscribed to the Hindu belief that the 'divine spark of God is present in all beings'. Baba devotee and biographer Professor N. Kasturi described him as a 'multi-faced avatar', claiming that he combined the life forces of the Hindu Rama, the Hindu Krishna, the Christian Christ, the Buddha and the 13th-century Iranian prophet Zoroaster.

Sai Baba's teachings fall into five loose, interconnected categories: right conduct, truth, love, peace and non-violence. His slogan seemed simple and appealing: 'Love All, Serve All'. When he died in 2011 the *Telegraph* newspaper obituary writer wrote that his appeal

> went far beyond the hippies and spiritual seekers who had made their way to India in the Sixties in search of enlightenment, that the numerous Sai groups which proliferated in Europe, America and Australia were liberally peopled with physicians, psychologists and teachers and that by the 1990s the tiny village of Puttaparthi had swollen to the size of a town and an airport was built to accommodate the growing numbers of pilgrims.[203]

Notable followers of Sai Baba include Narendra Modi, current prime minister of India, Indian Miss World winner and actor Aishwarya Rai Bachchan, and Isaac Tigrett, co-founder of the Hard Rock Café chain. Indie singer-songwriter Alanis Morissette even wrote a song about her visit to Sai Baba in 1998, though her lyrics are tinged with cynicism. Sai Baba was later surrounded by controversy; accusations by teenage boys of sexual abuse abounded, as did accusations of fraud .and there was even an apparent assassination attempt on him, in which his chauffeur and cook were killed – but in the years when Anna and Peter followed Sai Baba and visited him they were of course unaware of any possible basis for those accusations.

During the late eighties and early nineties the Roberts took three trips to visit the ashram of Sai Baba in Puttaparthi, northern India, travelling with

203 Accessed 21.6.16: http://www.telegraph.co.uk/news/obituaries/religion-obituaries/8471342/Sathya-Sai-Baba.html

two of their close friends, fellow devotees Betty and her husband Lorrie, who they'd met via the White Eagle Lodge. As you might imagine, ashram life was fairly regimented and ascetic: early rising for meditation, minimal conversing, a modest dress code (women were encouraged to wear sari), and lights off by 9 pm. Yet beyond all this focus on internal reflection and asceticism, occasionally there was a spot of magic and excitement, during one of these ashram visits Baba, magician-like, materialised a wristwatch. That same visit, during a private, small group meeting that Anna and Peter had been invited to, Baba had taken Peter aside and said to him, 'You will continue to do much for the animals, along with your good wife.' This, coming from the mouth of Baba at the time, was a meaningful statement, particularly for Anna, and the story of their private meeting with Baba and his 'prediction' was one she would often proudly repeat.

19

REWILDING

In early autumn 1991 the law of threes relating to bad news reared its head. First, Anna's brother Freddie, a year younger than her, died of lung cancer. He had never smoked, but he and his wife Jill had been the landlords of a pub for years while smoking in pubs and restaurants was still legal – and common. Then Anna's doctor told her that she had a sticky valve in her heart, and put her on Warfarin to thin her blood; she was very unhappy about this, preferring natural medicines and having known Warfarin only as an appallingly cruel rat poison.

Finally, the diagnosis came through that Peter had Parkinson's disease. After Freddie's funeral Anna and Peter came over to our family home in Petersfield and told Gillian. She was unsure what Parkinson's was, initially confusing it with Hodgkinson's, a form of lymphatic cancer. So she researched it, to find that it is a progressive neurological condition generally classified as a 'movement disorder'; the symptoms include tremors, slowed movement and bodily rigidity. Sufferers may shake uncontrollably, or freeze on trying to go through a doorway, or slow down and shuffle when they walk.[204] Other symptoms include loss of balance, fatigue and restless leg syndrome; in the later stages there can be a lack of bladder and bowel control. Parkinson's can cause communication difficulties such as slowed, slurred speech, memory impairment and trouble forming full sentences. Those living with Parkinson's have an increased chance of experiencing anxiety, depression, hallucinations.

The diagnosis had come about in large part because of the observations of a friend and former Compassion employee at my parents' wedding. After a decade-long engagement my parents, Gillian and Kevin, were married

---

204 http://www.newyorker.com/magazine/2014/04/28/have-you-lost-your-mind

by family friend and White Eagle Lodge minister and Compassion trustee Jeremy Hayward, in the round dome at Newlands. It was at the reception in the Copse House garden after the ceremony that some friends noticed that Peter seemed different – a little tired and distant – and that he stumbled occasionally. This was put down to exhaustion and age; he'd worked hard getting the garden and house perfect for his daughter's wedding. Nonetheless, friend and Compassion employee Jill Wright noticed a tremor in his hand, and herself having a family member with Parkinson's, she pushed Peter to go to the doctor for a check-up. Peter insisted that it was probably just a nerve thing, as he'd recently bumped his elbow on the car door – but Jill could see what it was.

Helen knew exactly what Parkinson's meant when they told her, as by that time she was working as a nurse at Basingstoke and Winchester Hospital. Anna and Peter sat her down in the lounge at Copse House and gave her the news; she reported that she felt 'surprised but not shocked'. Judy, Mike and their three daughters now lived in the Midlands – they had recently moved from Sunnyside Cottage to the Tall Barn in the village of Wootton, just outside Northampton – so Anna telephoned the news through. Just as Judy got off the phone from Anna, their new next-door neighbour Dave, a GP, popped over. Upset, she told him about the diagnosis.

'Of all the things you can get, that's one of the better ones,' he offered, kindly.

In the early days Anna went into denial; she and Jean le Fevre managed to convince themselves that the doctors had made a mistake. When it became obvious, however, that the doctors had not done so, Anna and Jean changed tack and predicted that Peter would be healed.

The doctors started him on a lowish dose of levodopa, the main drug used to treat Parkinson's. It is known as a miracle drug, one that revolutionised treatment in the sixties, and is still what neurologists call the 'gold standard'. It consists of a chemical building block which the body converts into dopamine, which helps with the symptoms, as dopamine loss is a major aspect of Parkinson's. Gillian supplemented her father's regime of pharmaceuticals with vitamins and supplements from The Bran Tub, as well as feeding him a dopamine-rich diet; he was quite content with this, eating tin after tin of Greek butter beans in tomato sauce, plus bananas and chocolate; and as well as food and pills there were massage, chiropractic and acupuncture. Lying supine, Peter suggested to Tammy, his acupuncturist and friend, that she project a copy of the Bhagavad Gita onto the ceiling so that her treatment wasn't so boring for him.

Peter was 67 at the time of diagnosis in 1991, and like many other sufferers he'd probably had the disease for a long time before that. He retired as CEO of Compassion that same year. At that point long-term employee and friend Joyce D'Silva, who had worked for Peter at Compassion since the mid-eighties, became the CEO of the charity. To both Anna and Peter, Joyce seemed a perfect choice. This, despite a bumpy start to her career there, in that back in the mid-eighties they had interviewed her for campaigns officer and had decided not to offer her the job. Later they explained to her that they had believed that her circumstances at the time were not conducive to the work; the job was based in Petersfield and was fairly high-pressure, but she lived in Essex and had three young children. This was of course before the internet and remote working, so the distance alone seemed like a serious stopper.

However, the person they had appointed for the job turned out to be such a terrible fit that they offered it to Joyce after all. She recalled Peter calling her at home one afternoon, when she had just got home from work. She was thrilled – and, selling her house and moving from Essex to Hampshire – started work as Compassion's campaigns officer in September 1985. Later she was promoted to campaigns director.

Joyce's consciousness around animals had already been well established when she joined the Compassion team; she had been vegetarian since 1971, a development she credits to reading Gandhi's autobiography, and then in 1975, after watching a BBC programme on dairy cows, she went vegan, a direction she'd already been moving in. Many of Joyce's friends at the time didn't quite get her change in diet and her consequent change in career – in fact, when she stopped eating meat her father thought she would die, and said to her husband Amancio, an Indian jazz musician, 'You mustn't let her do it.' But Amancio *did* let her do it – and later became vegan himself! As in the Roberts family, the legacy has continued: Joyce's two daughters are still vegan, as is her eldest granddaughter; her son is pescetarian, and his son (at the time of writing aged nine) is vegan too.

Joyce had grown up in Ireland, and prior to working for Compassion had graduated from Dublin University with a degree in Modern History and the History of Political Thought; she had taught Humanities in two schools in India and later in a comprehensive school in Essex, where she had become Head of RE. As with Anna and Peter, a deep interest in the scriptures of various faiths was a motivating factor for Joyce, and in Essex she had introduced a multi-faith syllabus, and established GCSE and A-level courses in RE.

Also like Anna, Joyce experimented in cookbook writing, and in 1980 produced the UK's second-ever vegan cookbook title, *Healthy Eating for the New Age*, published by Wildwood House – which ran to four editions!

In 1991 Joyce was appointed chief executive, a post she held for 14 years. She recalled that one of the major challenges during her time as CEO was

> getting [the] sentient beings petition (one million signatures) to the European parliament in 1991 and then seeing it through to 1997 when [they] got the Treaty of Amsterdam protocol. All [of which had been] started by Peter in 1988.

Joyce was a key figure in establishing the UK ban on sow stalls. The other major campaign of this era 'was the huge live exports campaign in 1995 – Peter [Stevenson], Philip [Lymbery] and John [Callaghan]' and Joyce, as she recollects, 'were running round in circles doing media interviews and organising demos at the ports'. She remains an ambassador for Compassion to this day.

For Peter's retirement the Compassion staff, by then around ten paid employees plus volunteers, bought him a professional-grade telescope. He planned to learn more about astronomy; he had always adored Patrick Moore's programmes and was fascinated with the celestial. Beyond star-gazing, his retirement included a great deal of gardening and continued intellectual pursuits; he never stopped reading and learning. He spent great swathes of time on the garden – as did Anna in the greenhouse, growing tomatoes, cucumbers and courgettes. Peter planted a heather-filled rockery plus a wall of red, yellow and purple tulips; dozens of huge pot plants overflowed across the patio. The garden was where the family elected to spend most of their time; aside from all the foliage there was the pool – and a jacuzzi! Although Peter enjoyed his premature retirement in planting, growing things, reading, taking family holidays in West Cornwall and the Mediterranean, I believe that had he not been afflicted by the myriad health issues caused in him by Parkinson's he would have died with his boots on.

Though at this point Peter officially retired as CEO of Compassion, he retained his input and remained on the board of directors for several more years, as did Anna.

Eventually, however, long board meetings became difficult for him to follow – and, frankly, to remain awake for – and so another milestone had to be faced. In 1999 the minutes of a meeting attended by Compassion in World Farming directors – chair Jeremy Hayward, welfarists Bill Jordan and Paul Rushton, and staff member Joyce D'Silva – read:

> The Directors note with considerable emotion a letter from Peter Roberts offering his resignation from both Boards. After so many years, it will be a sad loss to the Board. We also feel that the work of the CIWF group is moving into a global perspective

which was built into Peter's prophetic original vision and will continue to be led by Peter's example into the future.

By the time of Peter's full retirement from Compassion, he had been running the charity, or working for it in some capacity, for 32 years.

Peter was matter-of-fact about his health, and in the 15 Parkinson's years no-one I know ever recollected him complaining about his disease. He said to Gillian, 'These things happen,' and told her that she must make sure that 'one day, when I've gone gaga, you mustn't let me give everything away to Battersea Dogs' Home'! Kevin recalled that 'Peter just seemed to get on with it, not making any fuss' and that 'this period of illness only confirmed for him that Peter was a truly remarkable person'.

In these early years of the diagnosis, Helen and Peter were having tea together in the garden and talking about the condition. Someone asked Peter if he ever questioned 'Why him? And didn't he feel bitter?' He responded, 'Why *not* me?' This reads as if Peter was a stoic about his condition, but he had more humour about life than implied by the common perception of stoicism. For the last decade of his life he needed a stick, and at table would hook other people's glasses of wine towards himself. Then, on collecting an Animal Award from the BBC in the nineties, he asked the audience if 'anyone had seen this Parkinson's chap' because he 'wanted to have a word with him'.

After retiring as both CEO and board member, Peter was, until the last years of his life, still occasionally called at home for an opinion on something important by the next two Compassion CEOs, Joyce D'Silva and Philip Lymbery, and Peter and Anna would also attend the Compassion annual fundraising balls, usually held at the Dorchester Hotel in Park Lane, where they would be guests of honour. Supporters and patrons in attendance included actor Joanna Lumley, English singer Lynsey de Paul, actor Sue Jameson of the TV series When the Boat Comes In, and James Bolam of The Likely Lads, amongst others. Former Compassion employee John Callaghan recalls that 'Peter's courage through his years with Parkinson's was evident, as he continued to do all he could for his family and the animals right up to his last months'.[205]

In 1999, with Anna, Gillian and Kevin, he attended a talk by the Dalai Lama at the Royal Albert Hall: 'Ethics for a New Millennium'. At the interval, Joanna Lumley who was also in the audience, spotted Anna and Peter in the foyer; she was thrilled to see them, and emphasised her respect for all the good work they had done for animals and the planet.

205 John Callaghan, email, 6.8.17.

More awards followed, with an MBE in 2002 in recognition for his work on advancing farm animal welfare. He, humbly, had not wanted to be nominated for an MBE whilst he was still CEO, which was when Compassion (then) Campaigns Director Joyce had first suggested it. However, after he had retired Joyce decided to nominate him partly as a surprise, and partly as it would also raise the profile of Compassion's continuing work.

The nomination was accepted. A limousine picked up Peter and Anna, plus Judy, Gillian and Helen, from Copse House, to take them to Buckingham Palace to meet the Queen and receive the MBE scroll. As John Callaghan notes, 'It was great when Peter got his MBE for services to animal welfare after years of ridicule from farmers, vets and even other animal welfare groups.'

Interestingly, two decades prior to receiving the MBE, Peter had criticised the Royal household for their consumption of low-welfare animal produce. Journalist Richard North notes in his book *The Animals Report* how in 1982 'a royal warrant had been awarded to a firm which makes regular deliveries of battery eggs to Buckingham Palace.'[206] North then described how Compassion had begun sending the Queen two free-range eggs a day for her breakfast, along with some campaign material on egg production. Peter wrote in *Agscene* at the time how 'It's incredible that with 4,000 acres of royal farms that they are unable to supply her with non-battery-eggs.' The free-range eggs that Peter posted to the Queen had come from Bedales School's small farm; Princess Margaret's daughter, Lady Sarah Armstrong-Jones, was a pupil at the school.[207]

Beyond keeping his finger on the pulse of what was happening at Compassion as far as he could, in his retired years Peter also managed to recreate a semblance of his and Anna's early years as farmers – albeit in a vegetarian, not-for-profit hobbyist sense. In 1992 the Earl of Selborne, selling one of his fields, had offered local residents of Church Lane first refusal, which Peter had snapped up. He then obtained a council grant to plant hundreds of trees, bought a large red tractor and, with the help of some local forest workers, planted cherry, oak, ash and pine. Next, he and Anna dug a large pond, where they rehomed injured and rescued wild, ducks, geese and swans. Other than getting a local farmer to mow it for hay once every couple of years they left the trees and field to do their own thing, and it was soon teeming with wildlife.

A home movie from September 1993 shows that in the second year of the Roberts family enjoying the wild field it was a hot Indian summer. In the first frame Peter puts his tractor into gear and kangaroo-hops it across the gravel drive; his jumper matches the tractor. Anna comes into shot wearing a cerise sweater and pale denim jeans, holding her four-year-old grandchild Holly's

206 Richard North, *The Animals Report* (Penguin, Reading, 1983), p.44.

207 Richard North, *The Animals Report* (Penguin, Reading, 1983), pp.43–44.

hand. There is the loud purr of the machine in the background and a white noise of hysterical voices. The film pans across the field to the countless trees, still saplings encased in biodegradable plastic (to stop the deer destroying them). In the next shot there are nine of us family members on a garden bench in the back of the trailer, Zara the Alsatian at our feet, all being pulled along by Peter. He keeps turning back to check on us. We chug across the field. Next, Holly is on Peter's lap and seemingly in control of the wheel. Steam puffs out of the tractor's pipe like in a cartoon. Next we are leaping off the trailer straight into a haystack eight bales high. The camera pans to a shot of a dull orange, old-fashioned tent set up in the field; Anna and Peter pour out of it, with all their grandchildren bundled around them, followed by Zara, wagging her tail. Finally, we are at the bottom of the field, leaning against an old water butt, drinking some box-wine and tea. 'A refuelling stop. It's hard work isn't it, being a farmer?' says Kevin to Peter.

As time went on, however, the disease progressed and everyday activities became harder. Walking down the street took triple the time, and moving through The Bran Tub became an obstacle course. One technique Anna used for speeding Peter up was to count; this kept the momentum going and the distraction of the counting took away the worry of an impending fall. Petite Anna, then even more shrunk with osteoporosis, was a comic figure, shouting like a tiny sergeant major: 'Come on now Peter! *One-two-three-four, one-two-three-four*.' When he get really stuck, we'd push his knee from behind, lifting his leg into a walking posture, the only way to get him moving.

In the mid-nineties Anna and Peter holidayed with Helen in the south of France, and a couple of times Peter fell. One evening Helen recalls him getting upset. As a fiercely independent individual he said the thing that bothered him most was 'the thought of being dependent on others'.

Mostly, though, he seemed to maintain the positive outlook he'd always had and didn't appear to suffer from depression, although there were moments when one could read panic in his eyes. His sentences would fade away in the middle and he'd lose his point. Kevin remembers a time in the latter stages when the Parkinson's and the dementia were pretty bad; he was making conversation with Peter as best as he could when Peter seemed to suddenly snap right back into the moment, saying, 'Oh dear, it's getting worse isn't it?' Kevin tells how it was just a small moment but it seemed significant.

More and more frequently in the final years of his life Peter would cast his mind back 20 years or more, and he'd insist on going down to the Compassion offices to sort out some paperwork, send off some cheques, do something for the chickens. He didn't often hallucinate, but when he did see something it was usually an animal, often a sheep or cow. Rational explanations usually failed, and Anna, struggling, would call Gillian over. She would promise Peter that everything was fine at Compassion and that if there was anything extra to be done for the chickens then she would drive him to the offices the following day.

During his decline it was Anna and other members of the family who took on the majority of Peter's care, as Anna didn't like the ideas of strangers in her house, even if they were nurses and health care professionals. The only non-family member she would allow to help out was one of her favourite neighbours from Church Lane, who would come over and attempt to play board games with Peter.

Caring for Peter in these later years took its toll on Anna; she was herself then in her seventies with several ailments. Yet the strong between them remained strong till the end, and palpably so in the latter years. We would find them down at the bottom of the garden eating plums from their trees and listening to the birds; Anna would have managed to push Peter down in a wheelchair, despite her own frailty.

The autumn before Peter's passing, they celebrated their 50th wedding anniversary, and the whole family came to Copse House to celebrate their golden year. We decorated the garden with flares, strung up fairy lights, lit terracotta chimeneas, spread tealights across the tables, stuck long white dining candles into holders and floated lotus-shaped ones in bowls of water. We ate outdoors and drank lots of box wine. Nick, Helen's then husband, performed some hits from *Les Misérables* – one of Robert's favourite musicals – swaying back and forth, drink slopping, and singing its main protest song, 'Do you hear the people sing … when the beating of your heart echoes the beating of the drum?' This prompted us to gather buckets, wooden spoons, utensils, saucepans and colanders to drum on.

Nick toasted Anna and Peter on their 50 years together, and Peter presented Anna with a tourmaline and gold commemorative ring. At one point Mike took a photograph of Anna and Peter at the table watching the *Les Mis* drumming performance, laughing, looking up, wearing matching red fleeces, their faces lit up by all the candles. Later, Amy – Mike's daughter, and Anna and Peter's granddaughter – painted a portrait of them from the photo, and it hung over the fireplace in the Copse House lounge.

The following winter, one morning in November Gillian found Peter unconscious in bed and rang the emergency services; he was resuscitated and taken to Queen Elizabeth Hospital in Portsmouth. He'd had a mini-stroke. While Anna and the three daughters spoke to the doctors, I sat with him on the hospital bed. He was dressed in the standard NHS thin blue hospital gown, yet he began singing 'You are My Sunshine' to me – which, as he would put it, really turned on the waterworks!

One day he started wandering around the geriatric ward, the Lavender Ward, and fell, shattering his hip. He was given morphine and swiftly operated on, his hip set in a frog cast. Unfortunately, as is common when this happens to the elderly, his having to lie on his back, unable to move, meant that infection built up in his chest; shortly after the fall he developed pneumonia and was sent to the palliative care ward. At that time he asked that we, the family, 'shut off the lights'; clichéd as that sounds, he seemed content and ready to go.

After that he moved in and out of consciousness, speaking very little, and only in incomprehensible, whispery croaks. When I put my hand in his, there was no firm return grip, as there had been only a week before when one of the nurses had described him as 'as strong as a bull' (he'd kept pulling the tubes of antibiotics out of his arm and pushing the doctors away).

For the next three weeks Anna, Judy, Gillian and Helen only left his bedside in order to eat and sleep. A few days before Peter died Philip Lymbery visited him in the ward. Now middle-aged, Philip had worked for Peter with a couple of gaps since he was 25 years old, and in that last week when he came to say goodbye, Philip assured Peter that Compassion was in safe hands, and updated Peter about the latest animal welfare victories. It may have been wishful thinking but for a moment Peter seemed to understand, and he smiled.

He died on 15 November 2006.

Over a hundred mourners came to his funeral at Guildford Crematorium, the day after what would have been his and Anna's 51st wedding anniversary. Peter's coffin was covered in orange marigolds; orange was his favourite colour, his aesthetic stuck firmly in the seventies. Compassion trustee and lifelong friend Jeremy Hayward conducted the ceremony and there were so many people to seat that the Righteous Brothers' 'Unchained Melody' had to be looped five times. Jeremy spoke of Peter's legacy, his whirlwind love affair with Anna, their 50-year marriage, and Peter's love of family and the natural world. Jeremy quoted former Compassion employee John Callaghan's words: 'Peter had made it easy for us who followed him as he had made animal welfare campaigning respectable'.[208] We sung one of Peter's favourite hymns and William Blake's 'Jerusalem',

The reception was held at the Tithe Barn, an old converted stone sheep barn, the view outside one of verdant fields. The condolence cards overflowed the drawers of the dining room unit, usually full of biscuits and chocolate.

208 John Callaghan, personal correspondence, 6.8.17.

They included a card from Peter Singer, by now recognised as a major philosopher, acknowledging Peter's influence on him and his career. Almost all the major newspapers, tabloid to broadsheet, carried Peter's obituary, and in *The Economist* there was a double-page spread. The *Telegraph*'s obituary read,

> When they began, Robert's vision of a world in which farm animals were treated with respect and compassion looked a hopeless cause because they were mere agricultural products without any legal protection, and CIWF was seen as just another animal welfare campaign run on emotion and little else. However, Peter Roberts brought intellectual discipline to his cause and argued for the need for an agricultural system that would sustain the world.

'Wait for me, Peter,' wrote Anna on the card she put amongst the bright flowers at his funeral. They both believed in the afterlife, and when Peter died no one, least of all Anna, was expecting her to live another seven years.

But she did, even though after almost 51 years together she couldn't cope with the loss, and her physical and mental health declined massively. There was some happiness during these years, but much less, and after a while she moved into a mostly tranquillised state. She still loved family Christmases, watching the birds in her garden, reading the paper, and, later, watching football. The kitchen was perfect for wildlife watching and, as Anna often proudly claimed, it was 'really more of a wood than a garden'; a lack of distinction between Great Woods and their own garden meant the local wildlife – deer, pheasants, rabbits, badgers and foxes – would come very close to, and sometimes into, their house. Countless bird feeders meant the garden's birdlife was prolific, though Anna would become livid at the sight of squirrels inevitably 'stealing' the birds' food. Sometimes pheasants would wander into the house looking for extra scraps, and one afternoon a large rough-looking fox came into the kitchen, much to the shock of Anna's live-in carer, Mabel. It was so quiet at the bottom of the lane that outside the kitchen door in autumn the only sound was the leaves falling. This quiet suited Anna, though it didn't necessarily serve her in these later years as she became more reclusive after Peter's death and less inclined to entertain visitors beyond immediate family.

One happy memory from this time, however, was during a snowstorm in the January of 2010; my mother and father, brother Dom and his wife Emi, plus my best friend Charlotte (who Anna adored) and I, were all snowed in at

Copse House for almost a week. Dom and dad built an igloo on the patio and we took jacuzzis followed by rolls in the snow, with Anna watching out of the window. In the evenings, low on supplies, Gillian mixed tins of canned beans and tomatoes and whatever scraps of vegetables were left, and made a giant pot of stew. In the other saucepan we mixed an assortment of liqueurs found in the backs of the cupboards with fruit juice and red wine to make a winter punch that tasted like cough mixture. Anna generally declared that she hated snow (she worried about car accidents) but was clearly pleased to have us all trapped at her house.

Not long after this snowy enforced holiday, the Sunroom, the first headquarters of Compassion from all those years ago, took on its final and saddest incarnation as a makeshift bedroom for Anna, unable to make it up the stairs to her own room.

On 9 August 2013 she had a stroke, and within 24 hours of being taken to hospital she had died. Almost all the family, her daughters and grandchildren, were in the room when she took her last breath.

Her funeral mirrored Peter's. The Righteous Brothers looped once more, Jeremy conducted the service in his sky-blue Lodge robes, and his eulogy focused on Anna's desire to be with Peter again and her work for Compassion. He emphasised Anna's pivotal role in their decision to found Compassion, her belief that animals had rights and that something had to be done to assert them – and that she and Peter had done that something.

We held the reception in the garden of Copse House and there was some semblance of their 50th wedding anniversary celebrations from almost a decade earlier; everyone sat in a circle, flares in the grass, chimeneas lit, even though the tin one was now rusted through on one side.

The Christmas after she died was the year of massive storms. Trees had fallen during the night across the power lines of Hampshire and Sussex, and thousands of people spent a freezing Christmas in the dark and cold. We were able to heat the lounge and so stayed in it all together, huddled on the sofa around the open fire; the portrait granddaughter Amy had painted of them, based on Mike's photograph from their golden celebration, hung above the fireplace, their image watching us.

Today, sadly, despite widespread awareness of the issues Compassion has always campaigned on, the charity has far from 'put itself out of business' as Peter had always hoped it would. Compassion employee Phil Brooks notes how Peter assumed Compassion 'would be a temporary concern', but in spite of its great achievements over the past 50 years – the veal and sow crate ban, the battery hen ban, the 'sentient being' clause addition to the Treaty of Rome, and many other successes – there are currently more animals than at any other time in the history of the world suffering at the hands of factory farmers.

After receiving an animal welfare award from the British Veterinary Association in 1991, Peter gave a speech in which he concluded, 'If I sometimes appear discontented as regards the progress we have achieved together it is because it pales into insignificance when one sees the enormity of the task ahead' – and, more snappily, from a speech he gave at Compassion's 30th anniversary celebrations in 1997: 'There is still so much to do.'